Cultural Intelligence in Aviation

The aviation industry has evolved into a complex global network where diverse cultural backgrounds interact with flight operations, training, and safety management. *Cultural Intelligence in Aviation: Bridging Social Attitudes, Culture, and Safety in a Global Environment* examines the vital role of cultural intelligence (CQ) in creating safer, more effective aviation operations.

This book addresses the challenges and opportunities that arise when multinational crews, cross-border regulations, and varied workplace cultures come together in aviation environments. Through real-world case studies, empirical research, and theoretical frameworks, this guide explores how cultural intelligence shapes key aviation domains, from flight deck dynamics to accident investigations. It introduces innovative approaches to understanding cultural adaptability, incorporating methodologies from human factors, safety science, and artificial intelligence. Expert contributors bring a wealth of experience from across the industry and from different countries, allowing the reader to have a 360° view of the topic. Readers will gain practical tools and strategies to enhance cross-cultural communication, minimize miscommunication risks, and build a more inclusive aviation culture that prioritizes safety across all operations.

This resource is designed for aviation professionals at every level, including flight crews, safety managers, and training specialists. It will also serve as a valuable reference for academics and researchers in aviation studies, while providing regulatory authorities with evidence-based insights for policy development.

Cultural Intelligence in Aviation

Bridging Social Attitudes, Culture, and Safety in a Global Environment

Edited by
Dimitrios Ziakkas
Konstantinos Pechlivanis
Anastasios Plioutsias

CRC Press
Taylor & Francis Group
Boca Raton London New York

CRC Press is an imprint of the
Taylor & Francis Group, an **informa** business

Designed cover image: Shutterstock

First edition published 2026
by CRC Press
2385 NW Executive Center Drive, Suite 320, Boca Raton FL 33431

and by CRC Press
4 Park Square, Milton Park, Abingdon, Oxon, OX14 4RN

CRC Press is an imprint of Taylor & Francis Group, LLC

ISBN: 978-1-041-13362-9 (hbk)
ISBN: 978-1-041-13895-2 (pbk)
ISBN: 978-1-003-67202-9 (ebk)

DOI: 10.1201/9781003672029

Typeset in Times
by SPi Technologies India Pvt Ltd (Straive)

Contents

Preface ... x
About the Editors ... xii
Contributors ... xiii

Part I Foundations and Theoretical Context

Chapter 1 Rethinking Human Performance in High-Reliability Transportation Systems ... 3

1.1 Introduction ... 3
Anastasios Plioutsias, Dimitrios Ziakkas and Konstantinos Pechlivanis
1.2 Key Themes and Rationale ... 4
Anastasios Plioutsias, Dimitrios Ziakkas and Konstantinos Pechlivanis
1.3 Learning Outcomes and Audience Engagement ... 5
Anastasios Plioutsias
1.4 Book Structure and Roadmap ... 7
Anastasios Plioutsias, Dimitrios Ziakkas and Konstantinos Pechlivanis
1.5 How to Use This Book ... 9
Anastasios Plioutsias, Dimitrios Ziakkas and Konstantinos Pechlivanis
1.6 Conclusion ... 9
Dimitrios Ziakkas

Chapter 2 Aviation Safety, Human Factors, and Cultural Underpinnings ... 11

2.1 Introduction ... 11
Dimitrios Ziakkas and Hans C. Natakusuma
2.2 Traditional Models of Aviation Safety – The Role of Accident Causation Theories ... 12
Anastasios Plioutsias, Dimitrios Ziakkas and Konstantinos Pechlivanis
2.3 The Influence of Social Attitudes on Safety ... 16
Anastasios Plioutsias and Hans C. Natakusuma
2.4 Cultural Underpinnings in Aviation ... 17
Lea Sophie Trampitsch-Vink
2.5 Conclusion ... 23
Dimitrios Ziakkas and Anastasios Plioutsias

Chapter 3 Culture, Personality, and Identity in Aviation 26

3.1 Introduction 26
Dimitrios Ziakkas and Khalid Waseem
3.2 Theoretical Perspectives on Culture and Personality 26
Dimitrios Ziakkas, Khalid Waseem and Ian Roth
3.3 Religion, Coping, and Mental Health in Aviation 28
Dimitrios Ziakkas, Khalid Waseem and Hans C. Natakusuma
3.4 Cultural Scripts and Personal Identity in Aviation 31
Dimitrios Ziakkas, Khalid Waseem and Anastasia Kosmidou
3.5 Critical Reflections and Synthesis 33
Dimitrios Ziakkas, Konstantinos Pechlivanis and Khalid Waseem
3.6 Conclusion 35
Dimitrios Ziakkas, Konstantinos Pechlivanis and Khalid Waseem

Part II Cultural Intelligence and Crew Resource Management

Chapter 4 Understanding Cultural Intelligence (CQ) 39

4.1 Introduction 39
Anastasios Plioutsias, Dimitrios Ziakkas and Konstantinos Pechlivanis
4.2 The Four Dimensions of Cultural Intelligence (CQ)..............40
Anastasios Plioutsias, Dimitrios Ziakkas and Konstantinos Pechlivanis
4.3 Integrating CQ Dimensions in Aviation Context....................46
Anastasios Plioutsias, Konstantinos Pechlivanis and Dimitrios Ziakkas
4.4 Conclusion 47
Dimitrios Ziakkas

Chapter 5 Cultural Differences in CRM and Communication...........................49

5.1 Introduction 49
Dimitrios Ziakkas and Alex Pollitt
5.2 Multinational Crew Dynamics...51
Alex Pollitt and Dimitrios Ziakkas
5.3 Language and Communication Barriers.................................59
Spiros Chazapis
5.4 Communicating Across Cultures...62
Spiros Chazapis

5.5 The Organizational Culture: Safety Norms and Crew Resource Management Strategies 69
Daan Vlaskamp
5.6 Conclusion 74
Dimitrios Ziakkas

Chapter 6 Implementing Cultural Intelligence in Training 78

6.1 Introduction 78
Dimitrios Ziakkas
6.2 Developing Training Modules 78
Dimitrios Ziakkas, Anastasios Plioutsias, Dimitra Synodinou, Aspasia Argyrou and Ioanna Lekea
6.3 Debriefing Techniques 84
Dimitrios Ziakkas, Konstantinos Pechlivanis and Ian Roth
6.4 Case Vignettes and Examples 86
Dimitrios Ziakkas
6.5 Assessment and Feedback Mechanisms 88
Dimitrios Ziakkas, Konstantinos Pechlivanis and Hans C. Natakusuma
6.6 Conclusion 91
Dimitrios Ziakkas

Part III Accident Analysis Through a Cultural Lens

Chapter 7 Methodologies for Cultural Accident Analysis 95

7.1 Introduction 95
Dimitrios Ziakkas
7.2 Integrative Accident Investigation Frameworks 96
Dimitrios Ziakkas, Anastasios Plioutsias and Thanasis Binis
7.3 Proposed Framework 97
Konstantinos Pechlivanis and Dimitrios Ziakkas
7.4 Regulatory and Methodological Comparisons 98
Dimitrios Ziakkas, Thanasis Binis and Anastasios Plioutsias
7.5 Implementation of AI and Emerging Technologies 100
Dimitrios Ziakkas, Konstantina Ioannidi and Ian Roth
7.6 Conclusion 101
Dimitrios Ziakkas and Anastasios Plioutsias

Chapter 8 An Evidence-Based Approach 102

8.1 Introduction 102
Dimitrios Ziakkas

8.2 Recurrent Cultural Pitfalls in Aviation Safety 102
Anastasios Plioutsias, Odysseas Papaefthymiou and Yolanda Vaiopoulou
8.3 Pakistan International Airlines (PIA) Flight 8303 Case Study 104
Dimitrios Ziakkas, Anastasios Plioutsias and Konstantinos Pechlivanis
8.4 Tenerife Airport Disaster Case Study 107
Anastasios Plioutsias, Konstantinos Pechlivanis and Dimitrios Ziakkas
8.5 Synthesis of Findings and Implications for Future Practice 109
Anastasios Plioutsias, Konstantinos Pechlivanis and Dimitrios Ziakkas
8.6 Conclusion 111
Dimitrios Ziakkas

Part IV Strategies, Policy, and Future Directions

Chapter 9 Designing and Evaluating Cross-Cultural CRM Training Programs 115

9.1 Introduction 115
Dimitrios Ziakkas
9.2 Training Program Development 115
Dimitrios Ziakkas, Dimitra Synodinou and Ioanna Lekea
9.3 Proposed Framework – Evidence-Based Learning and Operational Simulations 117
Dimitrios Ziakkas and Hans C. Natakusuma
9.4 Evaluation and Feedback 120
Dimitrios Ziakkas and Ian Roth
9.5 Best Practices from Global Operators 123
Dimitrios Ziakkas
9.6 Conclusion 125
Dimitrios Ziakkas

Chapter 10 Policy, Organizational Change, and Future Trends 127

10.1 Introduction 127
Dimitrios Ziakkas
10.2 Policy Implications for Aviation Safety 128
Dimitrios Ziakkas and Anastasios Plioutsias
10.3 Leadership and Company Culture 130
Dimitrios Ziakkas and Konstantinos Pechlivanis

10.4 Implementing Cross-Cultural Teams and Feedback Systems 133
Dimitrios Ziakkas, Anastasios Plioutsias and Konstantinos Pechlivanis
10.5 Future Trends and Research Directions 134
Dimitrios Ziakkas, Anastasios Plioutsias and Konstantinos Pechlivanis
10.6 Conclusion 137
Dimitrios Ziakkas

Appendix A 140

Appendix B 145

Appendix C 148

Index 152

Preface

Aviation is, at its core, a human–technical ecosystem operating within tightly coupled, global socio-technical systems. As fleets, networks, crews, and aviation actors have become increasingly international, we have learned—often at cost—that safety cannot be secured by procedures and technology alone. It must also be understood through people: their cultures and identities, their languages and habits of communication, and the organizations that shape how they work. This volume takes that premise seriously. It offers a coherent program for reframing human performance, for strengthening investigation and training, and for institutionalizing cultural intelligence as a safety-critical competence across the aviation system.

The opening chapters revisit the conceptual ground on which modern aviation has been built. Rather than attributing adverse events to individual failings, they consider how people, technologies, and organizations interact in complex environments. Human variability—shaped by culture and context—is treated not as a liability to suppress, but as a resource for resilience when it is recognized, trained, and supported. In this sense, Just Culture is presented not as a slogan but as a practical management commitment that allocates accountability fairly while protecting learning.

Building on this foundation, the book broadens the lens through which competence is defined and assessed. The classical contributions of Crew Resource Management and organizational accident theory retain their value, yet the present era of automation, interdependence, and multicultural teams demands more. Communication fluency across languages and hierarchies, culturally informed situational awareness, and behavioral adaptability under time pressure are framed as positive capabilities that can be cultivated and evaluated within CBTA/EBT.

The middle chapters translate these ideas into practice. They examine how authority gradients, power distance, and independence–interdependence dynamics surface in everyday cockpit and cabin interactions, and they introduce Cultural Intelligence (CQ) as a trainable capacity with metacognitive, cognitive, motivational, and behavioral dimensions. Guidance is provided for embedding CQ systematically into training design—through scenario construction, guided role-play, and structured debriefing that produces observable behavioral change—so that culture becomes operational rather than rhetorical.

Investigation practice is addressed in depth. Traditional models such as HFACS, STAMP-CAST, and AcciMap remain useful, but they can miss culturally mediated behaviors unless investigators are trained to look for them. To that end, the book proposes an integrative method for identifying and coding cultural markers in evidence, for visualizing patterns, and for responsibly using digital twins and artificial intelligence to aid analysis. Case material—including well-known events such as PIA 8303 and Tenerife—illustrates how recurring cultural pitfalls can be translated into preventive strategies rather than post-hoc explanations.

The concluding chapters move from method to institutionalization. They outline what leaders must do to sustain change: maintain transparent reporting that protects voice, align incentives with learning, and ensure that policies, training, and oversight recognize intercultural competence as part of professional performance. Particular attention is given to the regulatory sphere. The volume is written to be useful not only to operators and training organizations, but also to regulators and Civil Aviation Authority directors who must evaluate AI-enabled systems, oversee performance-based programs, and steward data governance with public accountability. Templates for evaluation, examples of performance indicators, and suggestions for harmonizing assurance cases with ICAO, EASA, and FAA expectations are provided to support adoption. Emerging technologies—AI, digital twins, immersive simulation—are treated as facilitators of human judgment, not replacements for it.

Three claims thread the chapters together. Culture is operational, present in time-critical coordination as surely as in routine turnarounds. Competence is collective, dependent on how teams negotiate meaning, authority, and uncertainty across cultural and linguistic boundaries. Learning is designed, and organizations that institutionalize culturally informed training, investigation, and policy consistently outperform those that rely on tacit norms and good intentions. These claims are developed with the practitioner in mind, but they also speak directly to the needs of oversight authorities seeking credible, human-centered metrics for system performance.

The intended readership is broad: pilots, controllers, cabin crew, engineers and technicians; investigators, managers, and trainers; scholars of human factors and organizational behavior; and—importantly—regulators and CAA leaders charged with guiding the system's evolution. Readers will find conceptual clarity, empirical grounding, and practices that transfer: ways to embed CQ into CBTA/EBT, to code cultural variables during investigations, to evaluate cross-cultural CRM, and to align policy with organizational learning. The contributors neither romanticize culture nor pathologize it. They show how to harness it—through analysis, training, leadership, and fair oversight—to produce safer, more resilient operations. In that sense, the book is not only about culture in aviation; it is about the culture of aviation we choose to build.

We are indebted to the authors and reviewers for their rigor, and to the many practitioners who shared cases and insights. Our thanks extend to training centers, airlines, ANSPs, and authorities whose willingness to test culturally informed practices has made the work tangible. Any remaining shortcomings are my own. I offer this book as an invitation—to reconsider assumptions, refine methods, and redesign systems—so that human performance, in all its cultural richness, continues to advance safety in a global aviation ecosystem.

Dr. Dimitrios Ziakkas

About the Editors

Dimitrios Ziakkas is an affiliate researcher at the Science, Technology and Innovation in Society (STIS) Laboratory at the National and Kapodistrian University of Athens. He previously served as an Assistant Professor at Purdue University's School of Aviation and Transportation Technology, USA, and is an Airbus captain instructor and examiner with A-320, A-330/350, and A-340 type ratings. With over 34 years in military and civil aviation, he has accumulated 12,000 flight hours across various aircraft. A former Hellenic Air Force officer, he later served with Olympic Air, Qatar Airways, and Legend Airlines. He holds multiple advanced degrees, including a Ph.D. in History and Philosophy of Science and Technology, and focuses his research on aviation technology adoption, flight safety, and Advanced Air Mobility. He is a Fellow of the Royal Aeronautical Society and a member of several prestigious aviation organizations.

Konstantinos Pechlivanis is a an Airline Captain and expert in Human Factors. He holds a BSc in Aviation Science from the Hellenic Air Force Academy, an MSc in Human Factors in Aviation from Coventry University, and is currently pursuing a PhD at the Technical University of Crete. His scholarly interests focus on the integration of AI in aviation, Single Pilot Operations (SiPO), and competency-based frameworks for workforce planning. Konstantinos is an accredited Aviation Human Factors Specialist (AvHFS) by the European Association for Aviation Psychology (EAAP), Technical Member in the Chartered Institute of Ergonomics and Human Factors (CIEHF), Member of Royal Aeronautical Society (MRAeS) and Upper Freeman in the Honourable Company of Air Pilots.

Anastasios Plioutsias is Assistant Professor of Human Factors at Coventry University, UK. He has over ten years of experience in Human Factors and Accident and Incident Investigations. He was a fighter pilot in the Hellenic Air Force (HAF) for 27 years with flight experience in F-16BLK52+ and F-4E Phantoms (2500 flight hours in fighter). He is an experienced accident and incident investigator, and he has been assigned to all levels of investigation boards when he was on duty in HAF. He is an experienced accident and incident investigator and was assigned to all levels of investigation boards during his service with HAF.

Contributors

Aspasia Argyrou
National and Kapodistrian University of Athens
Athens, Greece

Thanasis Binis
HCAA
Athens, Greece

Spiros Chazapis
Human Factors Horizons
Bucharest, Romania

Ioannidi Konstantina
Hellenic Air Force
Athens, Greece

Anastasia Kosmidou
Human Factors Horizons
Bucharest, Romania

Ioanna Lekea
Hellenic Air Force Academy
Acharnes, Greece

Hans C. Natakusuma
Purdue University
Indiana, USA

Odysseas Papaefthymiou
ICAO SME HCAA
Athens, Greece

Konstantinos Pechlivanis
Technical University of Crete
Crete, Greece

Anastasios Plioutsias
Coventry University
Coventry, UK

Alex Pollitt
Cranfield University
Cranfield, UK

Ian Roth
Purdue University
Indiana, USA

Dimitra Synodinou
Human Factors Horizons
Athens, Greece

Lea Sophie Trampitsch-Vink
Austro Control
Vienna, Austria

Yolanda Vaiopoulou
Human Factors Horizons
Bucharest, Romania

Daan Vlaskamp
KLM
Amsterdam, Netherlands

Khalid Waseem
Purdue University
Indiana, USA

Dimitrios Ziakkas
National and Kapodistrian University of Athens
Athens, Greece

Part I

Foundations and Theoretical Context

1 Rethinking Human Performance in High-Reliability Transportation Systems

1.1 INTRODUCTION

Anastasios Plioutsias, Dimitrios Ziakkas and Konstantinos Pechlivanis

International aviation began in earnest following World War II, rapidly progressing from limited intercontinental flights to today's expansive global network. The establishment of organizations like the International Civil Aviation Organization (ICAO) in 1944 created foundational frameworks to manage and regulate international air travel. Throughout the second half of the 20th century, air transportation evolved dramatically. Advancements in aircraft technology, navigational systems, and regulatory harmonization led to unprecedented passenger and cargo growth.

As the industry expanded, airlines increasingly sourced crew members from various national backgrounds to manage growing international networks. Multinational crews quickly became the norm, particularly within carriers operating across continents or international alliances. New cultural dynamics were introduced into cockpit operations, cabin management, and ground-handling procedures, making intercultural interactions an operational standard rather than an exception.

Such diversity in the crew mix presents numerous advantages, notably, the capacity to cater effectively to diverse passenger demographics and the enrichment of organizational culture through diverse perspectives. Yet, it also introduces layers of complexity, especially when communication, decision-making, and teamwork depend significantly on cultural frameworks and understanding.

Globalization has interconnected nations and markets more closely than ever, intensifying cultural interactions within aviation. As the industry grew, operational challenges became closely intertwined with cultural dynamics. Differences in attitudes toward hierarchy, risk perception, decision-making authority, and operational protocols vary significantly across cultures, influencing team dynamics and Crew Resource Management (CRM).

Research consistently indicates that a crew composed of culturally diverse individuals encounters unique challenges and differing expectations around authority gradients and communication styles. Such discrepancies, when poorly managed, can impede clear communication, compromise situational awareness, and ultimately increase operational risks. Airlines and aviation authorities have recognized these

DOI: 10.1201/9781003672029-2

issues, progressively incorporating cultural awareness training into CRM programs. Despite these efforts, fully understanding and mitigating the complexities introduced by cultural diversity remain an ongoing challenge of utmost importance. Addressing these intricacies requires continuous, research-informed adjustments to training programs, standard operating procedures, and organizational culture to ensure both safety and efficiency in increasingly global aviation operations.

English was appointed as the standard language for aviation communication by ICAO to provide a common linguistic framework fundamental to international safety. Yet, significant challenges persist, primarily because most aviation professionals worldwide are non-native English speakers (non-NES). The disparity between NES and non-NES can inadvertently create communication barriers, causing critical misunderstandings. Even proficient non-native speakers can struggle with complex aviation phraseology, especially when accented speech, rapid delivery, or idiomatic expressions complicate interpretation. These language difficulties can have severe implications for safety – misunderstood clearances, overlooked instructions, or hesitations during critical operations.

The broader international aviation framework – characterized by global mobility, mixed-nationality crews, and variable language competencies – generates operational risks that are both subtle and systemic. Mitigating these risks depends on elevating cultural awareness and fostering adaptive communicative behavior. A methodical response to these challenges enables aviation professionals to maintain high safety margins while operating within increasingly complex multicultural settings.

1.2 KEY THEMES AND RATIONALE

Anastasios Plioutsias, Dimitrios Ziakkas and Konstantinos Pechlivanis

The aviation industry functions within a socio-technical ecosystem shaped by intersecting cultural norms, behavioral tendencies, and institutional structures. These factors influence not only interpersonal dynamics but also collective decision-making and organizational behavior. This complexity underscores the importance of expanding traditional frameworks, such as Cultural Intelligence (CQ), toward a more encompassing approach - Diversity Intelligence (DQ) - which provides a broader perspective on navigating intercultural dynamics.

Aviation decision-making (particularly under constrained timelines and elevated workload) often reflects underlying social attitudes. Among these, hierarchical orientation and cultural conservatism have demonstrated direct implications for communication flow, authority dynamics, and risk appraisal. In high power-distance settings, subordinates may refrain from questioning authority even in the face of procedural deviations or hazardous judgments. Historical case analyses have repeatedly shown how this reluctance to challenge seniority, even subtly, can contribute to escalation sequences culminating in loss-of-control or system failure.

Conservatism, by contrast, tends to manifest as institutional inertia. It impedes the uptake of novel safety mechanisms or revised procedures, even when evidence indicates their value. This attitudinal rigidity fosters adherence to legacy practices that may no longer reflect current operational risks. Organizations must counteract these tendencies with targeted interventions that promote open communication, reduce authority gradients, and enhance organizational willingness to embrace beneficial changes, thereby improving decision-making processes and safety outcomes.

The interplay between personality traits, cultural backgrounds, and leadership behaviors is especially salient in high-risk aviation environments. Certain personality attributes, such as conscientiousness, openness to experience, and emotional stability, are predictive of effective leadership in crisis situations. However, these traits do not function in isolation; rather, their manifestation is influenced significantly by cultural contexts. For example, leaders from collectivist cultures may emphasize consensus-building and harmony, whereas leaders from individualist cultures may prioritize decisiveness and assertiveness during critical incidents.

Understanding this intersection allows aviation organizations to tailor leadership development programs that consider both personality characteristics and cultural influences. Training that integrates these factors can enhance leaders' capability to manage diverse teams effectively, adapt to various cultural expectations, and perform optimally under pressure. This holistic approach not only improves team cohesion and decision-making but also ensures that leadership practices align effectively with the complex, dynamic demands inherent in aviation safety.

In summary, the transition from CQ to DQ, the recognition of how social attitudes influence decision-making, and the exploration of interactions between personality, culture, and leadership form key themes in understanding and addressing operational challenges in global aviation. These themes are not ancillary – they are central to any forward-looking approach to safety and performance. They establish the analytical foundation upon which subsequent chapters will elaborate, linking individual behavior, team function, and organizational outcomes in globally distributed and culturally diverse aviation systems.

1.3 LEARNING OUTCOMES AND AUDIENCE ENGAGEMENT

Anastasios Plioutsias

This section delineates the primary instructional aims and conceptual anchors that frame the content of the book *Cultural Intelligence in Aviation: Bridging Social Attitudes, Culture, and Safety in a Global Environment*. It serves a dual purpose: to establish a shared vocabulary for critical constructs and to orient readers to the mechanisms through which culture, communication, and safety intersect in modern aviation systems.

1.3.1 Defining Key Concepts

Four core concepts underpin the thematic progression of the text:

- Cultural Intelligence (CQ) refers to an individual's capacity to operate effectively in culturally diverse situations. CQ encompasses four dimensions – metacognitive, cognitive, motivational, and behavioral (Ang & Van Dyne, 2008; Livermore, 2011, 2024) – each representing a distinct set of intercultural competencies essential for global aviation teams.

- Diversity Intelligence (DQ) represents an extensive framework that transcends cultural boundaries to include various dimensions of diversity, including, but not limited to, gender, age, cognitive style, and socio-economic background. DQ enhances Cultural Intelligence (CQ) by considering the impact of intersecting social identities on team dynamics and safety behaviors (Hughes, 2023).
- Safety Culture refers to the collective organizational values, beliefs, and norms that influence safety-related attitudes and behaviors. A strong safety culture fosters transparency, advocates for the reporting of hazards, and encourages continuous learning and improvement (Reason, 1997; ICAO, 2018).
- Crew Resource Management (CRM) represents a structured approach to non-technical training, focusing on coordination, leadership, and interpersonal communication. Contemporary CRM incorporates cultural variability, making it particularly relevant for multinational or mixed-experience teams

(Helmreich, Merritt, & Wilhelm, 1999; Flin, O'Connor, & Crichton, 2008).

These constructs recur throughout the book and form the analytical lens through which case studies, policy critiques, and intervention models are interpreted.

1.3.2 Understanding the Impact of Social Attitudes

The social constructs significantly shape communication norms, authority gradients, conflict resolution methodologies, and responses to operational stress. For instance, individuals originating from high power-distance cultures may exhibit reluctance in questioning authority, which could potentially jeopardize aviation safety (Helmreich & Merritt, 1998; Hofstede, Hofstede, & Minkov, 2010). Comprehending the manifestation of these attitudes among diverse crews is imperative for mitigating risks associated with miscommunication, groupthink, or excessive deference to authority.

1.3.3 The Necessity of Culturally Adaptive Communication

In aviation, effective communication cannot be reduced to linguistic clarity alone. Rather, it must accommodate culturally contingent variables such as tone, gesture, spatial orientation, and conversational hierarchy. The following are operational strategies that enhance culturally attuned communication:

- Adjusting verbal tone and register in alignment with expected formality levels.
- Recognizing and interpreting culturally specific non-verbal signals.
- Employing explicit paraphrasing and closed-loop confirmation to verify mutual comprehension.
- Minimizing ambiguity through standardized phraseology while remaining sensitive to individual expression patterns.
- Reducing exclusionary patterns that arise from shared language or cultural shorthand within subgroups.

These practices are particularly pertinent during high workload, degraded communication conditions, or transitory phases of flight. When communication errors intersect with cultural misalignment, the risk of misinterpretation increases significantly. Studies in crew dynamics have shown that CQ-enhanced communication correlates with improved task allocation, workload sharing, and conflict management (Salas et al., 2006; Barshi & Farris, 2013).

The learning outcomes outlined above position the reader to critically engage with the subsequent material. They emphasize the necessity of an interdisciplinary understanding – combining psychological insight, sociolinguistic awareness, and operational expertise. This integrative approach reflects the demands placed on contemporary aviation professionals who must operate across cultural boundaries without compromising procedural compliance or safety margins.

Rather than prescribing fixed behaviors, this book advocates for situational competence: the ability to evaluate, adapt, and respond within diverse team settings under variable operational pressures. In this respect, CQ and DQ are not ends in themselves, but functional enablers of safety, reliability, and human performance in a globally distributed aviation ecosystem.

1.4 BOOK STRUCTURE AND ROADMAP

Anastasios Plioutsias, Dimitrios Ziakkas and Konstantinos Pechlivanis

The book is divided into four principal sections, each structured sequentially to reflect the dynamic interrelationship between culture, safety, and operational behavior within aviation contexts.

1.4.1 Part I – Foundations and Theoretical Context

The first part establishes the conceptual foundation of cultural intelligence (CQ), diversity intelligence (DQ), and the socio-cultural factors that influence aviation safety. It draws upon research in cross-cultural psychology, human factors, and organizational behavior to elucidate the issue: aviation operates within an increasingly globalized and culturally intricate environment where traditional safety frameworks must adapt (Helmreich & Merritt, 1998; Ang & Van Dyne, 2015; Livermore, 2024). Moreover, it examines the emergence of multinational crews, the impact of linguistic diversity on safety, and the socio-cultural attitudes that shape crew behavior and organizational culture. Readers are encouraged to engage comprehensively with these foundational concepts to cultivate the necessary conceptual scaffolding for understanding the remainder of the book.

1.4.2 Part II – Cultural Intelligence and Crew Resource Management (CRM)

The second part explores the integration of Cultural Intelligence (CQ) and Diversity Quotient (DQ) within Crew Resource Management (CRM) training and operational practices. Each chapter disaggregates a CQ dimension (i.e., metacognitive, cognitive, motivational, and behavioral) and links it to real-world aviation practices. Practical tools, including coding systems and scenario templates, are presented in alignment with current International Civil Aviation Organization (ICAO), Federal Aviation Administration (FAA), and European Union Aviation Safety Agency (EASA) guidelines (Flin, O'Connor, & Crichton, 2008; ICAO, 2018). The structure of this section facilitates progressive skill development, which is particularly beneficial for instructional design and organizational training programs. Educators and trainers may utilize this section as an independent manual for the enhancement of CRM courses, incorporating role-playing simulations, reflective exercises, and debriefing strategies that emphasize cultural adaptation and communication effectiveness (Salas et al., 2006).

1.4.3 Part III – Accident Analysis Through a Cultural Lens

The third part introduces a structured analytical model for investigating aviation incidents with attention to cultural variables. Drawing from the CQ framework, it presents a 20-item coding system designed to identify latent cultural contributors in flight crew behavior. This approach supports the integration of socio-cultural analysis into safety audits, occurrence reports, and accident investigation protocols (Reason, 1997; Ang et al., 2007). Safety investigators, regulators, and operational risk managers will find this section particularly applicable in identifying non-obvious precursors to system failure, especially in contexts involving multinational crews or culturally complex decision chains.

1.4.4 Part IV – Strategies, Policy, and Future Directions

The final part examines how cultural intelligence (CQ) can be systematically embedded into the policies, leadership practices, and training systems that shape the future of aviation safety. Drawing on real-world operational needs and global regulatory frameworks, it positions CQ not as an optional enhancement to Crew Resource Management (CRM), but as a central competency for modern aviation. The discussion begins with the design and evaluation of cross-cultural CRM training, showing how the ICAO ADDIE model and Kirkpatrick's evaluation framework can be adapted to build curricula that address cultural dynamics in communication, leadership, trust, and decision-making. The integration of CQ's metacognitive, cognitive, motivational, and behavioral dimensions is explored in depth, supported by advanced learning technologies, high-fidelity simulations, and AI-driven debriefing tools.

Policy sections highlight the need for ICAO, EASA, and FAA to codify intercultural competencies within safety regulations, ensuring that training explicitly addresses cultural adaptability alongside technical proficiency. Organizational change strategies stress leadership modeling, the creation of diversity-centric safety cultures, and the institutionalization of CQ within standard operating procedures, performance metrics, and crisis management protocols.

Future-oriented chapters examine how VR, AI, wearable biometrics, and interdisciplinary research can transform CQ training and assessment, creating adaptive, personalized learning pathways. These technological innovations are framed as tools to enhance – not replace – the human capacity for empathy, adaptability, and reflective judgment.

Ultimately, Part IV argues that the resilience and ethical integrity of global aviation will depend on how effectively it integrates cultural intelligence into every level of policy, leadership, and operational practice. In a sector defined by multinational crews and interconnected networks, the ability to bridge cultural boundaries is not a "soft skill," but a safety-critical requirement for the decades ahead.

1.5 HOW TO USE THIS BOOK

Anastasios Plioutsias, Dimitrios Ziakkas and Konstantinos Pechlivanis

This book is intended to fulfill various professional and academic functions, focusing on the following areas:

- As a Training Manual: Aviation trainers may incorporate CQ-based tools and simulations directly into CRM and human factors training programs. The book contains actionable templates, reflection prompts, and case studies.
- As a Course Supplement: University instructors specializing in aviation management, safety, or human factors may designate chapters as essential or supplementary reading. End-of-chapter discussion prompts and coding frameworks facilitate applied learning.
- As a Research Reference: Scholars and graduate students may utilize the conceptual models, literature reviews, and empirical frameworks presented throughout the book to inform theses, dissertations, and academic publications concerning culture and safety in aviation.

The modular design of the book enables selective use, while its conceptual integrity permits comprehensive coverage. Whether read linearly or consulted in parts, the material supports reflective inquiry and applied professional practice across a range of aviation safety domains.

1.6 CONCLUSION

Dimitrios Ziakkas

Aviation is not solely defined by its technology or systems architecture – it is fundamentally a human activity. Every checklist, standard callout, and procedural task is embedded within interactions among individuals who bring with them distinctive beliefs, communication habits, and expectations. These are not incidental. They shape how safety is practiced and how teams function under operational pressure.

As this opening chapter has argued, aviation personnel do not operate in a cultural vacuum. Authority gradients, decision latency, and crew dynamics are all modulated by the cultural frameworks individuals carry into the cockpit, the operations room, or the classroom. In environments where precision, synchrony, and trust are essential, these factors can either support or erode operational effectiveness.

This book does not propose a retreat from standardization. Instead, it argues for an expanded understanding of how procedural systems intersect with lived human variability. Cultural Intelligence (CQ) and Diversity Intelligence (DQ) offer structured, evidence-informed methods to identify and mitigate communication breakdowns, team friction, and risk concealment in culturally diverse contexts.

The following chapters offer both theoretical insight and practical implementation strategies. They examine training curricula, leadership behavior, and accident analysis through a cultural lens. The overarching objective is not uniformity but fluency: the ability to recognize, interpret, and act on cultural signals in ways that uphold safety and foster mutual understanding.

For aviation to sustain its reputation as a high-reliability sector, it must engage with the human dimensions that underpin its operations. Understanding those we fly with – and those we train, lead, and learn from – is not optional. It is operationally necessary

REFERENCES

Ang, S., & Van Dyne, L. (2008). Conceptualisation of cultural intelligence: Definition, distinctiveness, and nomological network. In S. Ang & L. Van Dyne (Eds.), *Handbook of cultural intelligence* (pp. 3–15). Routledge.

Ang, S., & Van Dyne, L. (2015). *Handbook of cultural intelligence* (2nd ed.). Routledge.

Ang, S., Van Dyne, L., Koh, C., Ng, K. Y., Templer, K. J., Tay, C., & Chandrasekar, N. A. (2007). Cultural intelligence: Its measurement and effects on cultural judgment and decision making, cultural adaptation and task performance. *Management and Organization Review*, 3(3), 335–371.

Barshi, I., & Farris, C. (2013). *Misunderstandings in ATC communication: Language, cognition, and experimental methodology*. Ashgate.

Flin, R., O'Connor, P., & Crichton, M. (2008). *Safety at the sharp end: A guide to non-technical skills*. Ashgate.

Helmreich, R. L., & Merritt, A. C. (1998). *Culture at work in aviation and medicine: National, organizational and professional influences*. Routledge.

Helmreich, R. L., Merritt, A. C., & Wilhelm, J. A. (1999). The evolution of Crew Resource Management training in commercial aviation. *International Journal of Aviation Psychology*, 9(1), 19–32.

Hofstede, G., Hofstede, G. J., & Minkov, M. (2010). *Cultures and organizations: Software of the mind* (3rd ed.). McGraw-Hill.

Hughes, C. (2023). *Diversity intelligence: Leading with inclusive insight in a complex world*. Leadership Press.

International Civil Aviation Organization (ICAO). (2018). *Safety management manual* (4th ed., Doc 9859). ICAO.

Livermore, D. A. (2011). *The cultural intelligence difference: Master the one skill you can't do without in today's global economy*. AMACOM.

Livermore, D. A. (2024). *Leading with cultural intelligence* (3rd ed.). HarperCollins Leadership.

Reason, J. (1997). *Managing the risks of organizational accidents*. Ashgate.

Salas, E., Wilson, K. A., Burke, C. S., & Wightman, D. C. (2006). Does crew resource management training work? An update, an extension, and some critical needs. *Human Factors*, 48(2), 392–412.

2 Aviation Safety, Human Factors, and Cultural Underpinnings

2.1 INTRODUCTION

Dimitrios Ziakkas and Hans C. Natakusuma

A rigorous understanding of aviation safety begins with a clear examination of how human performance is conceptualized, measured, and managed. For several decades, accident causation models have shaped not only how events are investigated but also how crews are trained and how safety itself is defined within organizations. Frameworks such as the Swiss Cheese Model, Reason's organizational accident theory, and the evolution of Crew Resource Management (CRM) contributed far more than terminology. They provided structured perspectives through which systemic vulnerabilities – often obscured beneath procedural compliance – could be exposed and addressed.

These models were introduced during periods of acute operational need. The Swiss Cheese Model illustrated how multiple small failures, seemingly unrelated, can align to produce catastrophic outcomes. Reason's work reframed the dominant narrative of pilot error by emphasizing latent organizational conditions and flawed system design. CRM, developed in response to cockpit breakdowns marked by communication failure, repositioned interpersonal coordination as a safeguard rather than a weak link.

Despite their continued relevance, the assumptions underpinning these frameworks require reassessment. Today's aviation operations are embedded within networks marked by automation, cultural heterogeneity, and increased interdependence. Safety outcomes are shaped as much by interpersonal dynamics, cultural expectations, and social perception as they are by procedural adherence or technical skill. Factors such as personality traits, group norms, and power-distance orientation influence how crews interpret authority, share workload, and respond to ambiguity.

This chapter does not seek to discard legacy models. Rather, it repositions them within the broader context of current operational realities. The goal is twofold: to reaffirm the enduring utility of established human factors theory while identifying areas where those models fall short – particularly in accounting for the cultural and nontechnical variables that now influence flight safety at every level. In this expanded framework, the emphasis shifts toward evidence-informed strategies that prioritize behavioral insight, cultural fluency, and the integration of nontechnical competencies. Competency-Based Training and Assessment (CBTA) emerges as a central

DOI: 10.1201/9781003672029-3

method for embedding these priorities into operational training, aligning evaluation with observable behaviors rather than generic role descriptions.

By building on prior theoretical foundations and incorporating new evidence from contemporary operations, this chapter sets the stage for a more comprehensive understanding of human performance in aviation – one that reflects the complex social and cultural architecture within which today's professionals operate.

2.2 TRADITIONAL MODELS OF AVIATION SAFETY – THE ROLE OF ACCIDENT CAUSATION THEORIES

Anastasios Plioutsias, Dimitrios Ziakkas and Konstantinos Pechlivanis

Historically, aviation safety has been conceptualized through various theoretical models that explain how accidents and incidents occur. Three influential frameworks have significantly shaped contemporary safety thinking: the Swiss Cheese Model, Reason's Model of accident causation, and the development of Crew Resource Management (CRM).

- **The Swiss Cheese Model**
  Introduced by James Reason in the early 1990s, the Swiss Cheese Model offers a layered representation of safety barriers, each containing inherent vulnerabilities – depicted as "holes" – which may momentarily align, allowing hazards to pass through unimpeded (Reason, 1990, 1997). These layers range from individual actions and frontline procedures to technological defenses and organizational controls. The strength of this model lies in its systemic orientation. It redirects focus away from isolated operator error and toward interactions between system-level conditions and active failures. Latent threats – such as design deficiencies, inconsistent training protocols, or competing operational demands – often remain dormant until triggered by frontline action. This approach reinforced the need for layered defenses and emphasized preemptive risk identification, long before safety management systems (SMS) became formalized.
- **Reason's Model of Accident Causation**
  Expanding on the Swiss Cheese concept, Reason's comprehensive accident causation framework identifies two primary types of failures: active failures and latent conditions (Reason, 1997). Active failures are immediate human actions or errors occurring at the operational frontline, typically evident in the immediate circumstances leading to an accident. Latent conditions, by contrast, are systemic inadequacies or vulnerabilities embedded within organizational structures, policies, and practices, often existing unnoticed for extended periods. These conditions, such as inadequate resource allocation, insufficient training programs, or flawed safety cultures, predispose the system to potential failures when activated by human error or unexpected events. This conceptual lens shifted investigative emphasis from surface-level procedural violations to systemic analysis. It invited scrutiny of organizational culture, leadership commitment to safety, and the distribution of

operational authority. Human error, in this framing, was not treated as the root cause but rather the final point in a causal chain already weakened by structural compromise.

- **Evolution of Crew Resource Management (CRM)**
  CRM emerged in the aftermath of accidents involving communication breakdown and poor crew coordination. Initially focused on the cockpit, CRM aimed to foster assertiveness, role clarity, and open communication irrespective of rank. It later broadened to encompass decision-making, conflict resolution, and task distribution within diverse teams (Helmreich, Merritt, & Wilhelm, 1999). Successive CRM iterations incorporated more granular behavioral competencies, embedding concepts such as mental model alignment, anticipatory leadership, and stress recognition. Modern CRM also acknowledges the impact of culture and social attitudes on team dynamics. Efforts to integrate cultural awareness into CRM curricula reflect the recognition that performance cannot be fully understood – or improved – without addressing underlying interpersonal and intercultural variables.

2.2.1 Human Factors in Aviation

Human factors represent the intersection of psychological, physiological, and social aspects that impact individual and team performance within aviation contexts. In this section, we will focus on the Nontechnical skills (NTS), which encompass communication, teamwork, decision-making, situational awareness, leadership, and workload management (Flin, O'Connor, & Crichton, 2008). These competencies are central to Evidence-Based Training (EBT)/Competency-Based Training Assessment (CBTA) approaches.

- **Communication and Teamwork**
  Effective Communication remains a core determinant of operational reliability. Failures in message transmission, clarity, and mutual comprehension can quickly escalate into unsafe conditions. Standardized phraseology alone does not resolve the challenges posed by multilingual crews or varying cultural norms around assertiveness and conflict. Open, rank-neutral dialogue must be deliberately cultivated through training and organizational reinforcement. Similarly, teamwork encompasses more than cooperation; it involves structured coordination, shared situational models, and a joint understanding of task priorities – particularly under stress. Effective teams rely on mutual monitoring, cross-checking, and feedback cycles that are sensitive to cultural dynamics yet grounded in operational precision.
- **Decision-Making**
  Aviation personnel routinely make high-stakes decisions under pressure. These decisions are shaped not only by cognitive processes but also by team interaction and cultural expectations. Deference to hierarchy, reluctance to challenge authority, or preference for consensus can all influence the quality and timing of decisions. NTS training aims to enhance situational risk assessment and structured decision pathways, reducing the influence of bias

and promoting situational adaptability. Integrating cultural insights into these training modules helps reconcile divergent approaches to command, challenge, and feedback across international crews.

- **Situational Awareness and Leadership**
  Situational awareness requires the continuous perception and interpretation of unfolding events and their implications. It is not static; it demands constant updating and recalibration. Breakdowns in awareness often stem from inattentional blindness, cognitive overload, or misaligned mental models within the crew. Leadership in aviation – both formal and functional – supports the preservation of awareness by structuring tasks, managing workload transitions, and intervening when errors compound. Effective leaders adapt their style based on the cultural composition of the crew and the nature of the operational context, balancing directive clarity with participatory openness

2.2.2 Cultural Considerations in Existing Safety Frameworks: A Comparative Review

Although most established aviation safety frameworks now acknowledge culture as a contributing factor, the way they operationalize it varies. Many rely on static, group-level constructs and do not integrate the more nuanced, individual-focused dimensions introduced by the Cultural Intelligence (CQ) model. The following analysis provides a comparative review of the role of culture in the creation and implementation of the existing safety frameworks worldwide.

- **Cultural Dimensions within the HFACS Framework**
  HFACS (Human Factors Analysis and Classification System) originally emphasized operational and organizational breakdowns. Later adaptations introduced cultural markers at the "organizational influences" level, referencing broad constructs such as "national culture" or "organizational norms." However, these references often remain generic and lack linkage to specific competencies. The model does not delineate between cultural awareness, adaptation, or motivation at the individual level. It fails to address, for example, how a pilot's metacognitive reflection or behavioral adjustment during intercultural interaction may influence error likelihood. In effect, the model recognizes culture as relevant but lacks granularity.
- **Helmreich and Merritt's Contributions on Culture and CRM**
  Helmreich and Merritt were among the first to empirically link national cultural traits to CRM outcomes. Their analysis – heavily influenced by Hofstede's dimensions – identified systematic patterns in how authority, collectivism, and uncertainty avoidance shaped cockpit dynamics. Their contribution was foundational yet bounded. They treated cultural orientation as largely fixed and group based. The CQ model, by contrast, conceptualizes cultural competence as trainable, flexible, and situated at the individual level. Bridging this gap remains an unresolved challenge in applied safety frameworks.

- **Cultural Modifications in Line Operations Safety Audit (LOSA)**
  Line Operations Safety Audits (LOSA) serve as a structured method for in-situ observation of flight crews, capturing threats, errors, and mitigation strategies in real time. Some LOSA implementations have been adapted to include cultural markers, noting breakdowns in communication or decision-making that appear culturally influenced-such as deference to authority, hesitation to speak up, or culturally specific misunderstandings. Yet these additions are often anecdotal or descriptive. Cultural phenomena may be documented qualitatively (e.g., "deference noted," "language barrier observed") but are rarely analyzed through a defined, CQ-based taxonomy. The absence of standardized indicators for each CQ dimension precludes systematic aggregation or comparison across cases. Even though they offer structured data collection that includes cultural context, there is an identified lack of operational coding aligned with the four-factor CQ model.
- **Cultural Ergonomics and Anthropo-technology in Other Domains**
  In other high-reliability domains (e.g., defense and healthcare), efforts grounded in anthropotechnology and cultural ergonomics have explored how human systems interact with localized cultural contexts. These models often employ ethnographic methods and focus on environmental adaptation. While valuable, they are rarely structured for integration into aviation's formal reporting, training, or auditing systems. Furthermore, they do not use CQ dimensions as organizing principles, limiting their interoperability with aviation-specific safety processes.
- **CQ-Focused Empirical Studies in Aviation**
  A growing body of empirical work has applied the Cultural Intelligence Scale (CQS) in aviation, particularly in examining pilot-controller communication, crew cohesion, and training efficacy. These studies demonstrate clear associations between CQ levels and operational behavior. However, integration of these findings into practice remains limited. No widely accepted model currently incorporates CQ metrics into risk classification, CRM evaluation, or event analysis workflows. The theoretical foundation is sound, but the operational application remains fragmented.
- **Synthesis and Future Considerations**
  Existing safety frameworks incorporate culture at a conceptual level, yet they do not systematically capture its role in individual behavior or decision-making. The absence of an incident analysis tool that explicitly maps CQ dimensions – Metacognitive, Cognitive, Motivational, Behavioral – represents a missed opportunity. A CQ-informed classification system would extend the analytical depth of tools like HFACS and LOSA by identifying specific competencies or deficits that influenced outcomes. Such a system would offer operational value by supporting both predictive assessments and post-event learning. Embedding CQ into CRM audits, training diagnostics, and reporting templates would not only improve cultural attunement but also enhance the precision of safety interventions. This integration is not a replacement for existing frameworks but a necessary augmentation – one that reflects the realities of multicultural operations and supports safer, more cohesive flight environments.

2.3 THE INFLUENCE OF SOCIAL ATTITUDES ON SAFETY

Anastasios Plioutsias and Hans C. Natakusuma

Aviation safety is often framed through procedural compliance, technical skill, and systemic oversight. Yet these are not the sole determinants of operational integrity. Social attitudes – rooted in ideology, personality, and cultural worldview – exert a measurable influence on how authority is interpreted, how risks are perceived, and how teams interact in high-reliability contexts. This subchapter rigorously examines the impact of Right-Wing Authoritarianism (RWA), Social Dominance Orientation (SDO), and established personality theories on aviation safety, particularly within multicultural and hierarchical contexts.

- **Right-Wing Authoritarianism and Conservatism**
  Right-Wing Authoritarianism (RWA) is a psychological construct characterized by submission to authority, adherence to conventional norms, and aggression toward individuals perceived as deviant or threatening (Altemeyer, 1996). Historically associated with conservative ideology, RWA emerges within social climates where stability, tradition, and obedience are prioritized over openness or dissent. In the field of aviation, pronounced RWA tendencies may reinforce steep authority gradients, particularly in culturally hierarchical environments. This phenomenon can inhibit junior crew members from voicing concerns, even when they identify critical errors (Helmreich & Merritt, 1998). Decision-making in such contexts may become deferential rather than collaborative, heightening the risk of unchecked errors in high-stakes operations. Conservatism, particularly in its social manifestation, often aligns with a preference for tradition, structure, and resistance to change. While these characteristics can foster discipline and order – qualities that are indispensable in aviation – they may also impede adaptive thinking and diverse viewpoints, both of which are critical for effective problem-solving and innovation in complex environments (Jost et al., 2003).
- **Social Dominance Orientation (SDO)**
  Social Dominance Orientation (SDO) reflects an individual's preference for hierarchy in social relationships and the dominance of higher-status groups over lower-status ones (Sidanius & Pratto, 1999). Within the field of aviation, where rank, nationality, or gender may engender implicit status hierarchies, SDO can subtly yet significantly influence interpersonal dynamics. Crew members exhibiting high levels of SDO are more inclined to accept inequality within the cockpit, resist inclusive practices, or diminish the contributions of lower-ranking colleagues. Such attitudes can undermine team synergy, diminish psychological safety, and inhibit the open exchange of concerns – particularly during cross-cultural operations (Chiu et al., 2013). Understanding SDO is especially pertinent when developing multicultural Crew Resource Management (CRM) training. Addressing implicit biases and fostering egalitarian team behaviors may assist in counterbalancing high-SDO tendencies and nurturing a more cooperative and inclusive environment.

- **Integration with Personality Theories**
  Research on personality offers valuable insights into how individuals navigate social and operational contexts within the aviation sector. Two predominant frameworks are particularly pertinent:

  1. Eysenck's Theory of Personality accentuates three dimensions: *extraversion–introversion, neuroticism–stability, and psychoticism* (Eysenck & Eysenck, 1985). For instance, individuals exhibiting high levels of neuroticism may demonstrate increased susceptibility to stress and errors during high-pressure situations, whereas those characterized by elevated psychoticism may display risk-prone or authoritarian behaviors.
  2. The Big Five Model (OCEAN) encompasses the dimensions of *openness, conscientiousness, extraversion, agreeableness, and neuroticism* (McCrae & Costa, 1997). High levels of conscientiousness are consistently associated with enhanced safety performance, while lower levels of agreeableness or openness may diminish team cooperation and cultural adaptability (Barrick & Mount, 1991).

Researchers and safety managers should exhibit a more profound understanding of individual differences in cockpit behavior. Understanding the interaction between social attitudes and personality traits supports more nuanced CRM interventions, refined leadership development, and predictive crew risk profiling. These variables should not be treated as abstract psychological curiosities – they directly inform how teams function under pressure, how authority is negotiated, and how effectively safety-critical information is communicated.

In practical terms, integrating RWA, SDO, and personality profiling into crew selection, line training, and debriefing processes can support more targeted behavioral feedback and risk mitigation strategies. The aim is not to pathologize individuals, but to enhance situational awareness about how attitudinal predispositions shape crew dynamics, communication flow, and ultimately, safety outcomes.

2.4 CULTURAL UNDERPINNINGS IN AVIATION

Lea Sophie Trampitsch-Vink

2.4.1 Defining Culture and Cultural Norms

The concept of "Culture," as a construct in psychology, human factors, and management, refers to the *shared beliefs, values, norms, and practices* that shape how individuals, teams, and organizations behave and make sense of the world (Hofstede, 1980; Schein, 2010).

Fundamentally, it can be defined as follows:

> "our shared beliefs, which predict our likely shared behaviours"
>
> (Vink & Walzl, 2025).

In aviation, culture determines how safety is prioritized, how communication occurs under stress, how authority is enacted, and how systems are designed and used. Culture is not just an academic and theoretical construct; it is the very environment in which human performance either thrives or fails and has repeatedly been shown to demonstrate and predict powerful outcomes, including human performance and system success (Kirwan & Shorrock, 2018).

The most widely used applied theory of *Culture* is conceptualized through Hofstede's (2001) framework, which includes the following cultural dimensions:

- Power Distance (acceptance of hierarchical order): Affects how readily crew members challenge authority or speak up.
- Individualism vs. Collectivism: Shapes decision-making and accountability structures.
- Uncertainty Avoidance: Influences attitudes toward procedures, adaptability, and flexibility.
- Masculinity vs. Femininity: Affects risk-taking, competition, and cooperation.
- Long-Term vs. Short-Term Orientation: Informs organizational investment in safety and human development.

These dimensions are particularly relevant in aviation, where operational safety depends on seamless collaboration among multicultural teams. In high-risk environments like the cockpit or the control tower, seemingly minor cultural mismatches can escalate into serious breakdowns of communication, coordination, or situational awareness (Helmreich & Merritt, 1998).

Indeed, Hofstede's theory is notably used by the International Civil Aviation Organization (ICAO), the Civil Aviation Navigation Service Organisation (CANSO), and many airlines, especially for Crew Resource Management training and safety investigations (ICAO, 2021a). There is strong predictive validity of the dimensions of Power Distance, Uncertainty Avoidance, and Individualism vs Collectivism, with the most important elements in aviation culture including Safety Culture, Just Culture, and Organizational Culture, which are themselves often part of national legislative frameworks (Reason, 1997). For example, high *power distance* is linked critically to safety behaviors and incident rates (Helmreich & Merritt, 1998), whereas a lower *uncertainty avoidance* has been mapped to Just Culture because it means generally more openness to learning from mistakes. Therefore, the understanding of Hofstede's Cultural Dimensions Theory is relevant in any discussion of culture within the aviation domain.

Beyond Hofstede's framework, two of the other most widely used cultural models in the aviation context are Gelfand's Tightness–Looseness Theory (2011) and Berry's Acculturation Model (1997). Gelfand's model offers a valuable lens for understanding how the strictness of social norms and tolerance for deviance influence organizational behavior and safety culture. In aviation, tight cultures (e.g., some Asian cultures) may exhibit strong compliance with procedures but may also discourage open reporting due to fear of sanctions, impacting the development of a Just Culture (Gelfand et al., 2011). Conversely, looser cultures may promote innovation and openness but struggle with consistent safety behaviors. Berry's Acculturation Model

(1997) is particularly relevant in the context of multicultural crews and international organizations, where employees must adapt to a dominant organizational culture while maintaining their cultural identity. For instance, expatriate air traffic controllers integrating into European ANSPs often navigate between assimilation and integration strategies, which can influence their communication, trust, and adherence to safety norms. Both models complement Hofstede's by emphasizing dynamic, contextual, and interpersonal aspects of cultural adaptation in high-reliability organizations.

Given the importance of these theories and the demonstrable links to wide-ranging aviation performance outcomes, the impact of culture cannot be overstated. Aviation is particularly sensitive to cultural variation due to its global, multilingual, and highly regulated nature. The same standard operating procedure may be interpreted differently depending on a crew's national or organizational culture. As such, understanding cultural norms is not peripheral but central to effective human factors practice, underpinning Crew Resource Management, Safety Management Systems, and organizational safety strategies.

2.4.2 Culture as the Foundation of Human Performance

One growing area where culture is seen as not just informative, but strategically and even tactically vital, is the construct of human performance. As outlined in Vink and Walzl (2025), human performance (HP) emerges from a socio-technical system shaped by four foundational layers: *culture*, infrastructure, talent management, and individual variability. The first and most influential layer: *culture* defines the operational atmosphere, the unspoken rules of behavior, and the mental models that guide safety-related decisions.

A strong, just, and transparent culture fosters performance-enhancing behaviors such as speaking up, peer monitoring, and error reporting (Reason, 1997; Hudson, 2001). Conversely, a punitive or opaque culture stifles communication, distorts data, and eventually erodes trust. Culture, in this sense, becomes the operating system for all human factors and, therefore, total human performance output.

Strictly speaking, ICAO defines safety culture as a set of enduring values and attitudes regarding safety, shared by every member of an organization (ICAO, 2021a). These shared beliefs influence everything from training effectiveness to fatigue management, and from decision-making in emergencies to leadership credibility.

In this sense, human factors are not purely biological or cognitive; they are also cultural. They are shaped by what the organization believes to be important and acceptable. Culture determines whether a fatigued controller is encouraged to speak up or silently pushed to "cope." It governs whether a near miss becomes a learning opportunity or a career-ending mistake (Vink & Walzl, 2025).

As Hudson (1999) succinctly put it: "*Culture is what people do when no one is watching*." And that makes it the most powerful predictor of latent risk in aviation systems.

2.4.3 National vs. Organizational Culture

It's important to distinguish between national culture and organizational culture, both of which influence aviation safety. National culture shapes deeply held beliefs acquired during socialization, while organizational culture is more malleable and defined by leadership, training, and shared operational experience.

For instance, Emirates and Lufthansa may operate similarly in procedural terms but demonstrate very different cultural approaches to authority, communication, and decision-making. National culture may drive an Emirati pilot to defer to hierarchy, while Lufthansa's training may emphasize assertive challenge, reflecting Germany's low power distance.

However, effective aviation organizations develop a robust organizational safety culture that can transcend national differences. This is achieved through standardized training (e.g., CRM), explicit values (e.g., Just Culture), and clearly defined escalation channels.

A striking example of how national culture can impact safety occurred in the 2013 crash of Asiana Airlines Flight 214 in San Francisco, where the aircraft struck the seawall short of the runway, resulting in three fatalities and over 180 injuries. The U.S. National Transportation Safety Board (NTSB, 2014) investigation found that the flight crew mismanaged the approach and failed to monitor the aircraft's automated systems effectively. Crucially, the report noted that the First Officer likely recognized the aircraft was in an incorrect mode but did not speak up assertively. This hesitation was attributed in part to cultural norms rooted in Korean society, particularly high power distance and deference to authority and seniority. The Captain, being older and more senior, may have inadvertently created a communication barrier that prevented timely intervention, and this was added to the fact that an additional check captain was present during the flight. This accident underscores how national cultural values can influence cockpit dynamics and highlights the importance of organizational strategies that promote assertive communication and flatten authority gradients, such as Crew Resource Management and Just Culture practices.

2.4.4 Cultural Influences on Communication and Behavior

2.4.4.1 Crew and Team Interactions

Communication errors are a major contributor to aviation incidents (Wiegmann & Shappell, 2003). The 1977 Tenerife disaster, the deadliest aviation accident in history, was in part due to culturally influenced miscommunication and power gradients in the cockpit (Helmreich, 1994). Since then, Crew Resource Management (CRM) has sought to neutralize these risks by flattening hierarchies and promoting open, assertive communication (Salas et al., 2006).

But culture still lingers. In high power distance cultures, junior staff may be reluctant to challenge a captain's decision, even if it violates procedure. This is known as *authority gradient*, and it remains a risk factor despite CRM training (Flin et al., 2008).

Cultural beliefs about seniority, shame, and group harmony all influence who speaks, when, and how. For example, collectivist cultures may avoid direct confrontation, choosing to imply or hint, which can lead to ambiguity in high-stakes environments (Matsumoto et al., 2001).

2.4.4.2 Conflict Resolution and Just Culture

Organizational culture influences how conflict is resolved – whether it is surfaced, suppressed, or weaponized. In a Just Culture, conflict is addressed through shared accountability and learning. In a Blame Culture, conflict is avoided or hidden for fear of punishment.

A Just Culture supports psychological safety, allowing individuals to report mistakes or near misses without fear of retribution (Dekker, 2007). This transparency improves learning and prevents incidents. In contrast, punitive environments erode trust and encourage silence, often until it is too late.

The European Union Aviation Safety Agency (EASA) and ICAO have both emphasized the importance of building Just Culture as a component of safety management systems (EASA, 2018; ICAO, 2021b). Yet, operationalizing this concept requires a profound cultural shift, not just policy change.

The Boeing 737 MAX accidents in Indonesia and Ethiopia are often cited as the tipping point in a long struggle for Boeing between more dominant business-driven cultures against safety cultures. Following two fatal crashes (Lion Air Flight 610 and Ethiopian Flight 302), investigations revealed a culture of silencing dissent, prioritizing schedule over transparency, and marginalizing engineering concerns (U.S. House Committee on Transportation and Infrastructure, 2020). Boeing's erosion of internal trust and its failure to integrate pilot feedback exemplify the catastrophic consequences of a collapsed safety culture (as cited by the Netflix Document "Downfall: The Case Against Boeing").

By contrast, the crew of Qantas Flight 32 (2010) is often celebrated for its exemplary CRM, safety culture, and dedication to open and transparent learning. Captain Richard de Crespigny and his team managed a catastrophic engine failure on an Airbus A380 and safely returned to Singapore with no fatalities despite almost being overwhelmed by system failure alarms and an unknown aircraft state, a true testament to training, assertive communication, and shared situational awareness (de Crespigny, 2012).

2.4.5 Risk Management and Cultural Beliefs

Cultural beliefs shape how individuals and organizations perceive and manage risk. Cultures high in uncertainty avoidance (e.g., Germany, Japan) tend to rely heavily on formal procedures, checklists, and audits. Cultures low in uncertainty avoidance (e.g., the United States, Australia) may tolerate more improvisation and flexibility (Hofstede, 2001).

These preferences influence everything from how checklists are completed to how deviations are reported. A robust risk management system, therefore, must account for these cultural predispositions, both at the national and organizational levels (Maurino et al., 1995).

Moreover, safety culture maturity, such as that described by Hudson's (2001) model, progresses from pathological ("who cares as long as we don't get caught") to generative ("safety is how we do business"). Moving up this ladder requires leadership commitment, system alignment, and most importantly, cultural intelligence.

Practically, many organizations embed Safety Culture into their Safety Management Systems. For example, in Europe, most Air Navigation Service Providers adopt the Eurocontrol model of Safety Culture which includes the elements of: a well-informed Culture, a strong Reporting Culture, a Just Culture, and a Flexible Culture, all leading to a culture oriented toward learning – the model is taken from Reason (1997).

2.4.6 Assessing and Shaping Culture

Culture is not fixed. It can be assessed, shaped, and improved. Several tools exist to support this, including:

- Safety Culture Surveys (e.g., ICAO's Safety Culture Toolbox, IATA's Safety Culture program)
- Human Performance Management (HPM) and Safety Management Standards (SMS) of Excellence assessments (e.g., CANSO SMS SoE and HPM SoE)
- Reporting Systems Analysis (e.g., EASA Just Culture Network)
- Various Organizational Culture surveys (such as the Organizational Culture Assessment Instrument (OCAI) of Cameron and Quinn (1999)).

These tools allow organizations to understand what their staff believe about safety, how comfortable they feel raising concerns, and whether policies align with day-to-day realities.

Leadership plays a critical role here. As noted by Guldenmund (2000), the espoused culture (what leaders say) must match the enacted culture (what actually happens). Discrepancies between the two erode credibility and prevent cultural maturity.

The goal is not to enforce uniformity but to cultivate alignment, where individual behaviors, team norms, and organizational policies reinforce one another in the pursuit of safety.

2.4.7 Culture as a Predictive System

Culture is not just explanatory. The goal of understanding and measuring it is to be able to attempt to predict behaviors. As our *Human Performance Pyramid* suggests, if we know the cultural baseline of an organization, we can forecast communication patterns, error likelihood, and organizational resilience (Vink & Walzl, 2025).

Safety metrics can be integrated with cultural indicators, like safety culture scores, trust levels, and reporting climates, into operational dashboards (Guldenmund, 2000). By quantifying these elements, we can build Human Performance Indices that reflect not just what happened, but why it happened, and what can be done to improve. Thus, culture is increasingly used in safety and operations as a predictive key-performance indicator (KPI).

2.4.8 Closing Thoughts

Culture is not an abstract concept. It is a design variable. It shapes everything in aviation, from how procedures are written, to how errors are managed, to how safety is understood. It forms the foundation of human performance and the boundary of what is considered "acceptable behavior." In today's complex, multicultural aviation environment, cultural intelligence is not just optional; it is essential, and more often than not, it is also required by laws and regulations. It enables organizations to predict human behavior, manage complexity, and design safer systems.

To achieve true safety and more efficient performance, we must constantly move beyond the cockpit, beyond the checklist, beyond the radar screen, and look into the beliefs that govern behavior. Because in aviation, shared beliefs always inform shared actions, and those actions determine whether a flight arrives safely or not at all.

2.5 CONCLUSION

Dimitrios Ziakkas and Anastasios Plioutsias

The traditional safety models that shaped modern aviation thinking (i.e., Reason's frameworks, CRM, and the broader human factors paradigm) have given us essential tools to see risk more systemically. They helped us move beyond blaming individuals, to understanding how errors emerge from the interplay of decisions, structures, and organizational blind spots.

Aviation today is no longer confined to shared assumptions, uniform crews, or standardized worldviews. Our flight decks are multicultural. Our teams are shaped by global norms, different languages, and diverse perceptions of authority, risk, and responsibility. The risks we now face are not just technical – they are relational, cultural, and human in the fullest sense of the word.

This chapter has shown that, while foundational models remain relevant, they must now operate alongside new ways of thinking. Cultural intelligence, individual variability, and evolving social attitudes are not footnotes to safety – they are central to it. Future safety frameworks must not only trace what went wrong but also understand *why* people acted the way they did, given who they are, where they come from, and how they were shaped.

If safety is ultimately a human outcome, then our models must reflect the full range of human experience – not just within procedures, but within the cultures, personalities, and beliefs that shape how people perform under pressure. To build the aviation systems of tomorrow, we must begin by fully seeing the people inside them today.

REFERENCES

Altemeyer, B. (1996). *The authoritarian spectre*. Harvard University Press.

Barrick, M. R., & Mount, M. K. (1991). The Big Five personality dimensions and job performance: A meta-analysis. *Personnel Psychology*, 44(1), 1–26.

Berry, J. W. (1997). Immigration, acculturation, and adaptation. *Applied Psychology*, 46(1), 5–34. https://doi.org/10.1111/j.1464-0597.1997.tb01087.x

Cameron, K. S., & Quinn, R. E. (1999). *Diagnosing and changing organizational culture: Based on the competing values framework*. Addison-Wesley.

Chiu, C. Y., Gelfand, M. J., Yamagishi, T., Shteynberg, G., & Wan, C. (2013). In search of multicultural competence in international business: Research on individual multicultural capacity and global organizations. *Journal of International Business Studies*, 44(6), 525–543.

de Crespigny, R. (2012). *QF32*. Penguin.

Dekker, S. (2007). *Just culture: Balancing safety and accountability*. Ashgate.

EASA. (2018). *Just culture guidance material*. European Union Aviation Safety Agency. https://www.easa.europa.eu

Eysenck, H. J., & Eysenck, M. W. (1985). *Personality and individual differences: A natural science approach*. Plenum Press.

Flin, R., O'Connor, P., & Crichton, M. (2008). *Safety at the sharp end: A guide to non-technical skills*. Ashgate.

Gelfand, M. J., Raver, J. L., Nishii, L. H., Leslie, L. M., Lun, J., Lim, B. C., … & Yamaguchi, S. (2011). Differences between tight and loose cultures: A 33-nation study. *Science*, 332(6033), 1100–1104. https://doi.org/10.1126/science.1197754

Guldenmund, F. W. (2000). The nature of safety culture: A review of theory and research. *Safety Science*, 34(1–3), 215–257. https://doi.org/10.1016/S0925-7535(00)00014-X

Helmreich, R. L., & Merritt, A. C. (1998). *Culture at work in aviation and medicine: National, organizational and professional influences*. Ashgate.

Helmreich, R. L., Merritt, A. C., & Wilhelm, J. A. (1999). The evolution of Crew Resource Management training in commercial aviation. *International Journal of Aviation Psychology*, 9(1), 19–32.

Hofstede, G. (1980). *Culture's consequences: International differences in work-related values*. Sage.

Hofstede, G. (2001). *Culture's consequences: Comparing values, behaviors, institutions, and organizations across nations* (2nd ed.). Sage.

Hudson, P. (1999). Safety culture - theory and practice. *Industrial Safety Management*, 27, 1–13.

Hudson, P. (2001). *A safety management maturity model*. Shell International Exploration and Production.

ICAO. (2021a). *Safety management manual (SMM)* (Doc 9859, 4th ed.). International Civil Aviation Organization.

ICAO. (2021b). *Safety culture: Guidance for operators and regulators*. International Civil Aviation Organization. https://www.icao.int

Jost, J. T., Glaser, J., Kruglanski, A. W., & Sulloway, F. J. (2003). Political conservatism as motivated social cognition. *Psychological Bulletin*, 129(3), 339–375.

Kirwan, B., & Shorrock, S. T. (2018). *Human factors and ergonomics in practice: Improving system performance and human well-being in the real world*. CRC Press.

Matsumoto, D., Yoo, S. H., & Nakagawa, S. (2001). Culture, emotion regulation, and adjustment. *Journal of Personality and Social Psychology*, 81(6), 921–936.

Maurino, D., Reason, J., Johnston, N., & Lee, R. (1995). *Beyond aviation human factors: Safety in high technology systems*. CRC Press.

McCrae, R. R., & Costa, P. T. Jr. (1997). Personality trait structure as a human universal. *American Psychologist*, 52(5), 509–516.

National Transportation Safety Board. (2014). Descent below visual glidepath and impact with seawall, Asiana Airlines Flight 214 (Accident Report NTSB/AAR-14/01). https://www.ntsb.gov/investigations/AccidentReports/Reports/AAR1401.pdf

Reason, J. (1990). *Human error*. Cambridge University Press.

Reason, J. (1997). *Managing the risks of organizational accidents*. Ashgate.

Salas, E., Wilson, K. A., Burke, C. S., & Wightman, D. C. (2006). Does crew resource management training work? An update, an extension, and some critical needs. *Human Factors*, 48(2), 392–412.

Schein, E. H. (2010). *Organizational culture and leadership* (4th ed.). Jossey-Bass.

Sidanius, J., & Pratto, F. (1999). *Social dominance: An intergroup theory of social hierarchy and oppression*. Cambridge University Press.

U.S. House Committee on Transportation and Infrastructure. (2020). *Final committee report: The design, development & certification of the Boeing 737 MAX*. U.S. Government Publishing Office.

Vink, L. S., & Walzl, A. (2025). *The human performance pyramid: A framework for understanding safety and complexity*. JustMinds Press.

Wiegmann, D. A., & Shappell, S. A. (2003). *A human error approach to aviation accident analysis: The human factors analysis and classification system*. Ashgate.

3 Culture, Personality, and Identity in Aviation

3.1 INTRODUCTION

Dimitrios Ziakkas and Khalid Waseem

Aviation operates at the intersection of human performance and high-risk technological systems, demanding not only precision and discipline, but also empathy, intercultural fluency, and adaptability. In such environments, individuals bring with them not just technical knowledge but deeply rooted personal, cultural, and social frameworks that shape their behavior and decision-making. Culture is not simply a background variable – it is part of the operating environment. This chapter explores the role of culture, personality, and identity in aviation, with a particular focus on how psychological traits and social scripts influence communication, leadership, and safety. The chapter also examines how regulatory frameworks – specifically those by ICAO, IATA, FAA, and EASA – approach these human factors in training, assessment, and operations, especially within the paradigm shift toward Evidence-Based Training (EBT) and Competency-Based Training and Assessment (CBTA).

3.2 THEORETICAL PERSPECTIVES ON CULTURE AND PERSONALITY

Dimitrios Ziakkas, Khalid Waseem and Ian Roth

Culture and personality in the aviation context draw from the traditions of psychological anthropology and cultural psychology. From the early efforts of Benedict (1934) and Mead (1950) to define "national character," the exploration of how collective patterns influence individual behavior has evolved into a sophisticated framework essential for understanding aviation safety, crew interaction, and cross-cultural performance.

While Benedict's and Mead's contributions have been rightly critiqued for essentializing culture and relying on anecdotal generalizations, their legacy lies in establishing the principle that culture is not merely an external system, but one that becomes embodied through habits, emotions, and thought processes. This understanding is foundational when interpreting behavior in high-risk, multicultural environments such as the cockpit, where swift decisions and trust between team members are imperative.

Contemporary developments in psychological anthropology, particularly the work of Markus and Kitayama (1991), underscore the relational construction of the self. Their theory of independent and interdependent self-construals has become particularly relevant in aviation contexts. Pilots trained and socialized in interdependent

DOI: 10.1201/9781003672029-4

cultures (e.g., East Asia, the Middle East, or parts of Latin America) often prioritize social harmony, group loyalty, and deference to hierarchy. These values, while adaptive in many social domains, can pose significant risks in flight operations when assertive challenge behavior is necessary. In contrast, pilots from more individualistic cultures (e.g., North America, Northern Europe, or Australia) may demonstrate greater comfort with direct communication and autonomy, sometimes at the expense of group consensus.

This divergence in cultural norms has concrete implications in Crew Resource Management (CRM) training and non-technical skills (NTS) development. One of the most pressing issues in global aviation is not only to foster effective communication, but to cultivate mutual cultural awareness. Training interventions must move beyond technical standardization and address deeply ingrained assumptions about authority, accountability, and emotional expression. For example, Helmreich and Merritt (1998) illustrated how pilots from high power distance cultures often hesitate to challenge captains, even when safety is at risk. Similarly, the Asiana Flight 214 accident in San Francisco (NTSB, 2014) revealed communication breakdowns influenced by cultural norms, demonstrating the continuing relevance of psychological anthropology in modern aviation safety management.

Institutional frameworks such as the ICAO Annexes 1 and 6, and EASA's Aircrew Regulation (Part-FCL), emphasize the integration of CRM and human factors into pilot training. However, the guidance often lacks granularity in terms of how cultural personality models are embedded into training content and assessment protocols. This is where a theoretical bridge to constructs such as Cultural Intelligence (CQ) becomes particularly valuable.

Moreover, psychological anthropology introduces a nuanced perspective by considering how individual agency operates within cultural scripts. The work of Shweder (1991) and more recent contributions in cognitive anthropology suggest that individuals actively negotiate their cultural contexts. Thus, attributing crew behavior solely to "national culture" may obscure the dynamic interplay between institutional training, personal identity, and situational pressures.

From a regulatory standpoint, ICAO promotes a harmonized approach to human factors through competency-based training and assessment (CBTA), formalized in the ICAO Doc 9995. The CBTA methodology echoes the insights of psychological anthropology by evaluating pilots not only on knowledge and skill, but on behavior in authentic operational contexts. Similarly, IATA's Evidence-Based Training (EBT) model aligns with the idea that behavior is shaped through complex cultural and psychological conditioning rather than procedural instruction alone.

Differences in regulatory emphasis are evident, however. EASA's approach tends to prioritize structured assessment frameworks, linking competencies to licensing standards (EASA, 2020), whereas the FAA places greater weight on scenario-based evaluation, drawing on pilot discretion and decision-making (FAA, 2023a). These differences reflect deeper institutional cultures – an aspect often neglected in regulatory discussions yet integral to psychological anthropology's attention to embedded systems of meaning.

In summary, psychological anthropology offers critical insights into aviation safety by contextualizing human behavior within cultural frameworks. It reminds us that behind every checklist and SOP lies a person shaped by social norms, emotional repertoires, and identity scripts. Integrating these perspectives into aviation training, particularly through CBTA and EBT methodologies, offers not only a more realistic portrait of human performance but a more humane and effective path toward safety enhancement.

3.3 RELIGION, COPING, AND MENTAL HEALTH IN AVIATION

Dimitrios Ziakkas, Khalid Waseem and Hans C. Natakusuma

Within the high-stakes world of aviation, where the error margin is minimal and performance is expected to remain consistently optimal, the psychological resilience of flight personnel is paramount. Despite rigorous selection, training, and recurrent evaluation protocols, human performance under duress remains a variable influenced by personal belief systems, coping mechanisms, and mental health. Among these influences, religion and spirituality occupy a subtle but potentially profound role in how aviation professionals process stress, maintain equilibrium, and navigate crises. Historically underexplored in aviation safety literature, religion warrants a deeper, integrative analysis that acknowledges its dual capacity to support psychological resilience or interfere with operational decision-making.

This subsection explores the intersection of religiosity, coping, and mental health in the aviation sector through a multidisciplinary lens. Drawing on psychological, anthropological, and aviation safety research, it evaluates the benefits and risks associated with spiritual belief systems in aviation contexts. Moreover, it provides real-world vignettes, regulatory perspectives, and evidence-based recommendations for integrating cultural and religious sensitivity into Crew Resource Management (CRM), Evidence-Based Training (EBT), and Competency-Based Training and Assessment (CBTA) programs.

3.3.1 Religiosity as a Coping Mechanism in High-Stress Environments

Religious belief systems often provide individuals with a cognitive framework through which meaning is assigned to suffering, uncertainty, and mortality – conditions not unfamiliar to aviation professionals. Pargament (1997) defines religious coping as the process through which individuals utilize their faith or spiritual beliefs to understand and deal with stressful life events. Within the cockpit, such coping may manifest as silent prayer, inner reliance on divine protection, or ritualized behaviors performed before or during flight.

Multiple studies have shown that positive religious coping – such as seeking spiritual support or reframing crises as growth opportunities – is associated with lower levels of anxiety and greater emotional recovery (Ano & Vasconcelles, 2005; Koenig, 2012). This is particularly relevant in aviation, where stress-inducing events such as engine failures, rapid decompression, or unpredictable weather may place crews under psychological pressure. For pilots and cabin crew members operating under prolonged circadian disruption and fatigue, faith-based coping can offer a stabilizing mental anchor.

However, not all religious coping is adaptive. Pargament et al. (1998) distinguish between positive and negative religious coping, the latter involving religious discontent, spiritual guilt, or passive deferral to divine will – sometimes referred to as religious fatalism. In an operational context, fatalism may discourage proactive problem-solving, fostering a sense of helplessness or over-reliance on metaphysical outcomes.

3.3.2 Peer Support Programs and Mental Health in Aviation

While physical and cognitive competencies are extensively assessed and monitored, the psychological wellbeing of pilots and crew has historically remained a more concealed issue, often shrouded in stigma and silence. The introduction and institutionalization of Peer Support Programs (PSPs) mark a paradigmatic shift – signaling a move toward proactive, compassionate, and human-centered strategies for managing mental health in aviation.

- **The Rationale for Peer Support**
  PSPs are built on the foundational principle that pilots and crew are more likely to disclose psychological distress and seek help when approached by a trained colleague rather than a formal authority figure. This peer-to-peer model, endorsed by EASA (2021) and ICAO (2022), aligns with the Just Culture framework, emphasizing open dialogue, trust, and non-punitive responses to vulnerability. EASA's *Regulation (EU) 2018/1042* mandates that all commercial air operators under its jurisdiction implement a support program that includes psychological assessment and peer support mechanisms. The regulation was a direct response to high-profile incidents such as the Germanwings Flight 9525 tragedy, which underscored the devastating impact of unrecognized and untreated mental health conditions (BEA, 2016).

  Effective PSPs rely on several key components:

- **Voluntary Access**: Participation is typically voluntary, preserving individual agency and privacy.
- **Confidentiality**: Conversations between the peer supporter and the individual are protected by strict confidentiality protocols, fostering openness and psychological safety.
- **Training of Peer Supporters**: Selected personnel are trained in basic counseling techniques, active listening, and the escalation process for severe cases.
- **Referral Pathways**: When necessary, PSPs connect individuals with professional mental health services, ensuring a continuum of care.

  Many PSPs also operate in coordination with Aeromedical Examiners (AMEs), who are increasingly encouraged to adopt a facilitative rather than solely evaluative role (ICAO, 2021; EASA, 2022).
- **Cultural and Operational Considerations**
  Mental health interventions in aviation cannot succeed without accounting for cultural attitudes toward psychological distress. In some regions or

populations, discussing emotional vulnerability may be perceived as weakness or professional inadequacy. As a result, PSPs must be culturally attuned – training peer supporters to recognize not only psychological symptoms but also the social scripts that may prevent open dialogue. Operational demands – such as irregular schedules, long-haul fatigue, isolation, and circadian disruption – further compound the risks of mental strain. PSPs help buffer these stressors by fostering a sense of community, normalizing stress responses, and offering real-time debriefing after critical incidents.

- **Efficacy and Evidence**
  Empirical research supports the efficacy of peer-led interventions in mitigating stress, enhancing coping, and reducing burnout. Beernaert et al. (2021) maintained that over 80% of pilots who accessed European PSP programs reported improved psychological resilience and operational performance. Similarly, the FAA (2023b) has recently expanded its support for mental health screening, citing a reduction in grounded pilot time when early interventions are employed through PSPs.

3.3.3 Mental Health, Religion, and Cultural Humility

Mental health in aviation is gaining overdue attention, particularly following high-profile incidents linked to psychological distress. While regulatory bodies such as EASA and ICAO have begun advocating for peer support programs and mental health disclosure protections, the cultural and religious contexts in which individuals interpret mental health remain unevenly addressed (EASA, 2018; ICAO, 2021).

For many aviation professionals, particularly those from collectivist or religiously conservative backgrounds, traditional Western approaches to mental health – such as individual psychotherapy – may carry social stigma. Religious institutions, by contrast, may offer community, purpose, and accessible pathways to emotional expression and recovery (Koenig, 2012). Recognizing this dynamic requires a shift toward cultural humility in mental health programming, integrating religious literacy and optional faith-based support into wellness initiatives.

Aviation operators with diverse crews must be particularly attuned to these dimensions. A pilot from a Muslim-majority country, for example, may turn to an imam or engage in ritual prayer for comfort before seeking psychological counseling or peer support program. Religious flight attendants may view fasting as a form of spiritual discipline that intersects with emotional regulation. Understanding such practices allows training managers and airline psychologists to develop inclusive strategies that respect personal belief systems while safeguarding operational standards.

3.3.4 Integration into EBT and CBTA

Evidence-Based Training (EBT) and Competency-Based Training and Assessment (CBTA) frameworks emphasize observable behaviors and context-specific skill application. Incorporating cultural and religious coping styles into these frameworks presents a challenge – but also an opportunity – for modern aviation training. For instance, while spiritual coping may not be an assessed competency, its impact on behavior during high-stress situations is significant. One proposed approach is to include modules on personal belief systems and stress management within the resilience-building components of EBT curricula. These modules could provide crews with tools to recognize when religious beliefs are supporting versus inhibiting performance and offer CRM-compatible strategies for integrating personal values into team dynamics without compromising safety.

Additionally, simulator training scenarios may incorporate culturally relevant stressors – such as operating during religious holidays, fasting schedules, or cross-cultural crew pairings with varying religious sensitivities. These scenarios would not test religious belief but would prepare crews to navigate diversity with empathy and professionalism – an increasingly vital skill in globalized aviation. Religion and spirituality remain underexplored yet vital components of psychological coping in aviation. As the sector continues to diversify and expand globally, integrating cultural and religious literacy into training, mental health programs, and safety investigations becomes essential. Far from being antithetical to rational operation, faith – when harmonized with evidence-based procedures – can serve as a resource for emotional stability, meaning-making, and resilience. Future aviation systems must adopt a more inclusive understanding of mental health, one that honors the full spectrum of human experience, belief, and identity in the cockpit and beyond.

3.4 CULTURAL SCRIPTS AND PERSONAL IDENTITY IN AVIATION

Dimitrios Ziakkas, Khalid Waseem and Anastasia Kosmidou

This section provides an in-depth analysis of how cultural scripts shape personal and professional identity in aviation. It situates this discussion within the broader frameworks of psychological anthropology and human factors science and seeks to critically assess how diverse identities manifest in Crew Resource Management (CRM) scenarios, decision-making under pressure, and Evidence-Based Training (EBT) environments. The chapter also provides regulatory comparisons among ICAO, IATA, FAA, and EASA and offers practical pathways for integrating identity awareness into Competency-Based Training and Assessment (CBTA).

- **Cultural Scripts: A Theoretical Framework**
  Cultural scripts are shared cognitive models that guide behavior in social settings. These scripts are acquired through language, rituals, and lived experience and are internalized as implicit instructions for how to behave, speak, or react in given scenarios (Wierzbicka, 1994). In the aviation sector, cultural scripts influence how professionals interpret hierarchy, respond to conflict, express emotion, and negotiate risk. These patterns are not trivial – they can either strengthen crew coordination or silently erode situational awareness.

  For instance, in high power-distance cultures such as those in East Asia or Latin America, pilots may feel it is inappropriate to openly challenge a superior – even in the face of evident procedural deviations. In contrast, individuals from low power-distance cultures like Scandinavia or New

Zealand are more likely to view assertive disagreement as a mark of professionalism. These scripts can have profound implications for CRM and flight safety.

- **Identity and Authority Gradients in Aviation**
  Professional identity in aviation is strongly linked to notions of authority, leadership, and discipline. Yet these are interpreted differently across cultural contexts. In some collectivist societies, respect for seniority and harmony within teams overrides personal initiative. In these cases, first officers might withhold dissent, not out of ignorance or fear, but out of deep cultural conditioning. The notion of authority gradients – steep or shallow – captures how power and voice are distributed within cockpit teams (Helmreich & Merritt, 1998). Studies have shown that fatal accidents have, in some cases, been attributed to steep authority gradients where junior crew failed to intervene effectively.
- **Intersectionality: Gender, Ethnicity, and Role Expectation**
  While much of the aviation literature focuses on national culture, identity in aviation is also shaped by gender, ethnicity, age, and socioeconomic background. A female captain in a patriarchal society may face both overt and subtle resistance. Pilots from underrepresented ethnic backgrounds may feel additional pressure to prove competence, leading to over-cautious or over-assertive behaviors. Cultural expectations can influence help-seeking behavior, willingness to disclose fatigue, and communication styles. These facets of identity become visible during stressful operational phases – go-arounds, diversions, or CRM disagreements. Identity is not a variable to be "controlled for" but a central axis of human performance.
- **EBT and CBTA: Training Toward Identity Awareness**
  Evidence-Based Training (EBT), as supported by ICAO Doc 9995, offers a flexible, performance-oriented framework that can accommodate cultural and identity-related variability. Instead of solely measuring adherence to checklists, EBT emphasizes observable competencies. ICAO (2022) has underscored the importance of assessing crew members not only on technical knowledge but on their adaptability, communication strategies, and emotional regulation.

To embed identity-awareness, training programs must:

1. Include scenario-based debriefs that explore how identity influenced perception and behavior.
2. Foster open dialogue about cultural assumptions.
3. Develop trainers' cultural intelligence to support psychologically safe learning environments.

- **Regulatory Perspectives: ICAO, EASA, FAA, and IATA**
  - ICAO: Emphasizes the global harmonization of human performance principles, including the cultural dimensions of safety management (ICAO, 2022).

- EASA: Focuses on psychological competencies and social behaviors under CBTA frameworks. EASA has integrated cultural awareness into Part-FCL training requirements (EASA, 2021).
- FAA: While technically rigorous, the FAA still lags in integrating cultural dimensions into CRM and training programs. Its emphasis remains largely on procedural compliance, though CRM is mandatory for Part 121 operators.
- IATA: Has acknowledged the role of cultural diversity in training and supports global harmonization through EBT and IOSA standards. IATA's 2023 Human Factors guidelines include identity-relevant competencies such as inclusivity, assertiveness, and active listening.

Aviation is a technical domain, but it is also a deeply human enterprise. Cultural scripts and identity must be recognized not as complications, but as core realities of operational safety. Integrating these into training, regulation, and organizational culture requires humility, academic rigor, and a commitment to inclusion.

3.5 CRITICAL REFLECTIONS AND SYNTHESIS

Dimitrios Ziakkas, Konstantinos Pechlivanis and Khalid Waseem

As we move from theoretical foundations and cultural narratives toward reflective synthesis, we are faced with an enduring challenge in aviation psychology: to what extent are professional behaviors culturally conditioned, and to what degree can individuals modify their conduct considering new experiences, structured training, and introspective insight? The tension between cultural determinism and personal agency is not merely philosophical. It has immediate consequences for how safety is understood, how competencies are evaluated, and how crews are trained across diverse regulatory and operational contexts.

3.5.1 Cultural Determinism and the Question of Human Attitude

The concept of cultural determinism posits that individual behavior is largely shaped – if not predetermined – by social and cultural norms. For example, in aviation, many behaviors observed in multicultural cockpit environments, such as deference to authority or reluctance to challenge procedures, can be interpreted through this lens (Hofstede, 2001; Helmreich & Merritt, 1998). Yet the counterargument, rooted in contemporary psychological anthropology and human factors research, emphasizes the power of reflexivity – individuals, even within tight cultural scripts, retain the capacity for change, adaptation, and self-directed learning (Markus & Kitayama, 1991).

In the training context, this debate translates into practical concerns: Can a pilot from a high power-distance culture be effectively trained to assert concerns in a hierarchical cockpit environment? Can training truly penetrate deep-seated cultural scripts without undermining individual identity? Evidence from aviation training programs suggests that agency can be cultivated, particularly when training design emphasizes context-sensitive communication, self-awareness, and scenario-based reflection (Salas et al., 2006).

- **CBTA and EBT: Structured Pathways for Behavioral Adaptation**
  Competency-Based Training and Assessment (CBTA) and Evidence-Based Training (EBT) serve as the operational platforms through which agency and cultural awareness can be systematically addressed. These frameworks mark a departure from legacy models based on flight hours or rigid procedural knowledge. Instead, they focus on observable behaviors and decision-making competencies demonstrated under realistic operational conditions. This transition is supported by global authorities, including ICAO (Doc 9995), EASA (Part-FCL), and IATA's IOSA framework.

  CBTA identifies key core competencies that reflect both technical and non-technical performance dimensions. These include:

- **Application of Procedures**: Demonstrated understanding and adherence to SOPs across normal and non-normal situations.
- **Communication**: Use of clear, timely, and assertive communication that respects cultural and linguistic diversity.
- **Situational Awareness**: Ability to accurately perceive, interpret, and project future states under dynamic conditions.
- **Problem-Solving and Decision-Making**: Capacity to evaluate options and execute safe, effective solutions under pressure.
- **Leadership and Teamwork**: Skills in coordinating crew behavior, resolving conflict, and encouraging participation.
- **Workload Management**: Competency in managing task priorities, especially under time pressure or high demand.
- **Cultural and Human Performance Consideration**: While not always explicitly listed, many advanced CBTA programs now include modules on cultural awareness and interpersonal variability (EASA, 2021).
- **EBT: Beyond Standardization Toward Adaptive Expertise**
  EBT complements CBTA by leveraging data analytics and threat-based scenario construction to tailor training to actual operational risk. Instead of assuming one-size-fits-all curricula, EBT enables instructors to adjust for performance trends across different operational environments. This makes EBT particularly suited to address cultural variables, as patterns in behavior can be correlated with background variables, leading to deeper root cause identification and training customization.

  For example, if a training program identifies that assertiveness is consistently lower in communication tasks among crew members from specific cultural backgrounds, instructional design can shift to focus on role-playing assertive interventions, rephrasing techniques, and team dynamics. These methods are particularly effective when paired with structured debriefing tools that allow crews to reflect on their behaviors and their cultural roots.

  (Flin and Agnew, 2018)

3.5.2 Reframing Competency Through a Cultural Lens

It is essential to recognize that the very idea of "competency" is culturally constructed. What constitutes effective communication or leadership in one culture may be interpreted differently in another. A pilot trained in North America may value directness and initiative, while another trained in East Asia may prioritize group harmony and deference to rank. Rather than imposing a monocultural model of excellence, CBTA and EBT frameworks must account for this variation.

Recent trends in EASA and ICAO guidance encourage incorporating cultural dimensions into the assessment of non-technical skills. For example, leadership might be demonstrated not through overt authority, but through emotional regulation, empathetic listening, or silent coordination – behaviors often undervalued in Western training paradigms but critical in collectivist cultures (Gelfand et al., 2011).

3.6 CONCLUSION

Dimitrios Ziakkas, Konstantinos Pechlivanis and Khalid Waseem

The cultural and psychological dimensions of aviation safety cannot be overlooked. As this chapter demonstrates, flight safety is not just a matter of checklists and procedures but a product of human interpretation, emotional regulation, and social navigation. Culture influences how information is shared, how authority is enacted, and how decisions are made. Personality shapes how individuals react to stress, ambiguity, and responsibility. Religious beliefs and mental frameworks affect how individuals find resilience and meaning.

Modern training frameworks such as CBTA and EBT, supported by ICAO, IATA, FAA, and EASA, provide the structural flexibility to integrate these human elements. Cultural Intelligence, personality insight, and mental wellbeing must now be viewed not as ancillary training topics but as operational competencies. To move aviation safety forward, we must abandon the binary of cultural determinism vs. personal attitude. Both perspectives are valid, and both are incomplete when considered in isolation.

Culture provides the scaffolding for perception and behavior, while the new approach enables transformation within and beyond those boundaries. It is at their intersection that effective, inclusive, and adaptive aviation training emerges. CBTA and EBT provide the structure for this transformation. When paired with reflective instruction, inclusive performance metrics, and culturally literate leadership, they become not just tools for safety but for human development. In the end, safety in aviation is a shared, negotiated practice – not simply between pilot and aircraft, or between procedure and performance – but between the human mind and the social world that shapes it. The future of aviation safety depends on understanding not only the system but the human beings who operate it.

REFERENCES

Ano, G. G., & Vasconcelles, E. B. (2005). Religious coping and psychological adjustment to stress: A meta-analysis. *Journal of Clinical Psychology*, 61(4), 461–480.

BEA. (2016). Final Report on the accident on 24 March 2015 to the Airbus A320-211 registered D-AIPX operated by Germanwings. https://www.bea.aero/en/investigation-reports/notified-events/detail/accident-to-the-airbus-a320-registered-d-aipx-operated-by-germanwings-on-24-03-2015/

Beernaert, N., Van Dessel, P., & De Vooght, K. (2021). Peer support in aviation: Evaluation of European programs. *Journal of Aviation Psychology and Applied Human Factors*, 11(1), 42–53. https://doi.org/10.1027/2192-0923/a000173

Benedict, R. (1934). *Patterns of culture*. Houghton Mifflin.

EASA. (2018). *Just culture: Guidance material*. European Union Aviation Safety Agency.

EASA. (2020). *Acceptable means of compliance and guidance material to Part-FCL*. European Union Aviation Safety Agency.

EASA. (2021). *Guidelines for competency-based training and assessment (CBTA)*. European Union Aviation Safety Agency.

EASA. (2022). Safety promotion leaflet: Peer support programmes. https://www.easa.europa.eu/en/downloads/134715/en

FAA. (2023a). *Aviation instructor's handbook (FAA-H-8083-9C)*. U.S. Department of Transportation, Federal Aviation Administration.

FAA. (2023b). Enhancing pilot mental health through peer engagement strategies. https://www.faa.gov/newsroom/faa-mental-health-initiative

Flin, R., & Agnew, C. (2018). *Human factors in safety management: safety culture, safety leadership and non-technical skills*. CRC Press.

Gelfand, M. J., Raver, J. L., Nishii, L. H., Leslie, L. M., Lun, J., Lim, B. C., … & Yamaguchi, S. (2011). Differences between tight and loose cultures: A 33-nation study. *Science*, 332(6033), 1100–1104. https://doi.org/10.1126/science.1197754

Helmreich, R. L., & Merritt, A. C. (1998). *Culture at work in aviation and medicine: National, organizational and professional influences*. Ashgate.

Hofstede, G. (2001). *Culture's consequences: Comparing values, behaviors, institutions and organizations across nations* (2nd ed.). Sage.

IATA. (2023). *Human factors and safety management manual*. IATA.

ICAO. (2021). *Manual on mental health promotion in aviation (Doc 10110)*. International Civil Aviation Organization.

ICAO. (2022). *Manual of evidence-based training (Doc 9995)*. International Civil Aviation Organization.

Koenig, H. G. (2012). Religion, spirituality, and health: The research and clinical implications. *ISRN Psychiatry*, 2012, 1–33.

Markus, H. R., & Kitayama, S. (1991). Culture and the self: Implications for cognition, emotion, and motivation. *Psychological Review*, 98(2), 224–253.

Mead, M. (1950). *Male and female: A study of the sexes in a changing world*. William Morrow.

National Transportation Safety Board. (2014). *Descent below visual glidepath and impact with seawall, Asiana Airlines Flight 214* (Accident Report NTSB/AAR-14/01).

Pargament, K. I. (1997). *The psychology of religion and coping: Theory, research, practice*. Guilford Press.

Pargament, K. I., Smith, B. W., Koenig, H. G., & Perez, L. (1998). Patterns of positive and negative religious coping with major life stressors. *Journal for the Scientific Study of Religion*, 37(4), 710–724.

Salas, E., Wilson, K. A., Burke, C. S., & Wightman, D. C. (2006). Does crew resource management training work? An update, an extension, and some critical needs. *Human Factors*, 48(2), 392–412.

Shweder, R. A. (1991). *Thinking through cultures: Expeditions in cultural psychology*. Harvard University Press.

Wierzbicka, A. (1994). Cultural scripts: A new approach to the study of cross-cultural communication. *Applied Linguistics*, 15(1), 1–23.

Part II

Cultural Intelligence and Crew Resource Management

4 Understanding Cultural Intelligence (CQ)

4.1 INTRODUCTION

Anastasios Plioutsias, Dimitrios Ziakkas and Konstantinos Pechlivanis

The increasingly global nature of aviation operations necessitates a profound understanding of intercultural interactions. Crew members are regularly required to navigate diverse professional settings, characterized by cultural, ethnic, generational, and cognitive differences. Historically, Cultural Intelligence (CQ) has provided a comprehensive framework for addressing these cross-cultural dynamics. CQ has been extensively studied and widely adopted in global contexts, defined as an individual's capability to function effectively in culturally diverse environments (Livermore, 2011, 2024). CQ encompasses a set of competencies that enable professionals to interpret and manage cross-cultural situations effectively, emphasizing adaptation, communication, and cultural awareness. Although CQ has provided substantial improvements in intercultural competence, contemporary aviation environments demand an even more extensive framework, addressing diversity not only through cultural differences but also through dimensions such as gender, age, cognitive styles, educational background, and socioeconomic status. Recent scholars have argued for the expansion of CQ toward Diversity Intelligence (DQ), offering a broader and more nuanced approach for aviation safety and efficiency (Hughes, 2023).

DQ broadens the scope of CQ by explicitly addressing these multiple dimensions of diversity. Hughes (2023) defines DQ as the capability of recognizing, valuing, and effectively managing diversity in organizational settings. Aviation demands professionals who can integrate this multifaceted diversity into coherent operational practices. Thus, transitioning from a CQ framework toward DQ involves cultivating awareness and skills beyond cultural boundaries, systematically addressing how various forms of diversity impact crew interactions, leadership effectiveness, and ultimately, safety and performance outcomes.

This chapter introduces DQ as a functional evolution of CQ in aviation contexts. It examines how extending the scope of intercultural competence to encompass structural and identity-based diversity enhances crew coordination and operational safety. By articulating the mechanisms through which DQ can be embedded in aviation systems, this chapter aims to provide both conceptual clarity and applied relevance to practitioners, educators, and regulators operating in the global aviation ecosystem.

DOI: 10.1201/9781003672029-6

4.2 THE FOUR DIMENSIONS OF CULTURAL INTELLIGENCE (CQ)

Anastasios Plioutsias, Dimitrios Ziakkas and Konstantinos Pechlivanis

Aviation professionals operate within complex, multicultural ecosystems where communication, coordination, and decision-making unfold under pressure. CQ provides a functional model for understanding and managing these interactions. Defined as the capacity to function effectively across diverse cultural contexts (Ang & Van Dyne, 2008; Livermore, 2011, 2024), CQ comprises four interrelated dimensions. These are metacognitive, cognitive, motivational, and behavioral. When contextualized within the broader framework of DQ, these dimensions serve as operational tools for navigating interpersonal variability in high-reliability environments.

Integrating CQ and DQ within aviation involves embedding these competencies into Crew Resource Management (CRM) training, leadership development programs, and organizational safety cultures. Effective CRM training modules increasingly incorporate CQ and DQ principles, preparing crews to anticipate and manage intercultural dynamics proactively. Exercises that simulate rank deference, indirect refusal, or implicit disagreement provide trainees with the opportunity to identify, decode, and respond to culturally coded behaviors. These activities should not aim to "correct" cultural differences, but to expand interpretive flexibility.

Leadership training that integrates CQ and DQ equips aviation professionals to navigate role expectations and group dynamics more effectively. This includes developing reflexes for inclusive communication, recognizing unspoken resistance, and managing interpersonal uncertainty. Instructional design should incorporate elements of cultural pattern recognition, but also foster individualized strategies for assertiveness, empathy, and task alignment.

At the organizational level, policy alignment is essential. Safety culture frameworks must incorporate CQ and DQ principles explicitly – not only in training documentation but in behavioral audits, recruitment strategies, and peer evaluation systems. This includes establishing reporting structures that account for cultural silence, reluctance to escalate, or indirect conflict expression. Without institutional reinforcement, individual competence remains under-leveraged.

Transitioning from CQ to DQ marks a conceptual reorientation. While CQ emphasizes culture-specific variation, DQ expands this view to include identity categories and intersectional variability (e.g., cognitive diversity, generational norms, and educational conditioning). In practice, this shift recognizes that communication failure or decision inertia often reflects overlapping identity constraints, not merely cultural misunderstanding.

The annexes provide structured material that complements the main text by detailing the Cultural Intelligence (CQ) codes, their mapping to pilot non-technical skills, and examples of training modules and simulations. Appendix A introduces the four CQ dimensions—Metacognitive, Cognitive, Motivational, and Behavioral—each broken down into codes (e.g., MCQ1: Pre-interaction planning, CCQ3: Knowledge of nonverbal cues, MQC2: Persistence and confidence, BCQ4: Role flexibility). These are illustrated with practical indicators and aviation-specific examples.

Appendix B presents their integration into five core pilot competencies (communication, workload management, situational awareness, problem-solving, and leadership/teamwork), effectively linking cultural intelligence to existing Crew Resource Management (CRM) frameworks. Appendix C extends this with full CRM training modules and simulation templates, including case studies (e.g., Korean Air CRM transformation, Avianca Flight 52), role-play exercises, VR scenarios, and structured debriefings.

The decision to place these materials in the annexes rather than the main body reflects both editorial economy and reader usability. Including full code tables and workshop templates in the chapters would disrupt the narrative flow and risk overwhelming readers with excessive technical detail. By positioning them in annexes, the book preserves a coherent progression of argument and discussion in the main chapters, while still making the applied resources readily available. This separation also recognizes the dual readership: academics benefit from a smooth theoretical and empirical discussion in the text, whereas CRM instructors and human factors practitioners can consult the annexes directly for practical implementation without having to extract dispersed details from the core chapters. This balance respects the limited "book estate" while ensuring the content remains actionable for training and operational contexts.

In summary, the four dimensions of CQ provide a scaffold for understanding how crew members perceive, process, and respond to cultural variability. When integrated within a broader DQ framework, these dimensions become tools not just for interpersonal adaptation but for systemic improvement. Applied effectively, they enhance safety, sharpen decision-making, and strengthen cohesion in diverse operational teams.

4.2.1 Metacognitive CQ: Awareness and Reflection

Metacognitive CQ pertains to the capacity to reflect on and regulate one's cognitive processes during intercultural interactions (Ang & Van Dyne, 2008). It encompasses the preparation for cultural exchanges, the mindfulness exercised during those exchanges, and the reflection that occurs afterward. This "thinking about thinking" distinguishes routine interaction from culturally attuned engagement. Rather than functioning on habit or default scripts, individuals high in metacognitive CQ engage in structured thought processes that promote adaptation and reduce ambiguity in culturally heterogeneous settings. In aviation, where communication precision and behavioral alignment are critical, metacognitive CQ offers a performance advantage. Crew members who consistently assess cultural cues, test assumptions, and modify their communicative approach in real time contribute to stronger coordination, reduced misinterpretation, and higher levels of psychological safety. The deliberate regulation of interaction transforms standard CRM execution into culturally responsive teamwork.

Metacognitive cultural intelligence encompasses three primary processes: planning, monitoring, and reflection. Planning involves deliberate preparation for anticipated cultural differences. This may include reviewing regional SOP adaptations, noting cultural norms that affect authority dynamics, or adjusting phrasing and tone in pre-flight briefings. A captain preparing for a multinational flight deck may assess likely interactional patterns and recalibrate command phrasing to pre-empt confusion or silence. Monitoring refers to the ability to track interactions as they unfold. Individuals engage in real-time adjustments by noticing hesitations, seeking clarification, or modifying tone to match the moment. This internal self-observation, paired with external cue recognition, forms a critical feedback loop in dynamic team settings. Reflection encompasses post-interaction analysis – evaluating what was understood, what failed to connect, and how adjustments can be made in future settings. It supports long-term improvement by embedding learning into procedural memory. This may involve structured journaling, peer feedback solicitation, or critical review of cockpit audio and team performance data. These processes establish a continuous learning loop that fosters cultural adaptability over time (Thomas et al., 2008).

High levels of metacognitive CQ have been associated with more adaptive, inclusive, and strategically effective performance. Individuals who develop this capacity tend to notice subtle dissonance early, preempt escalation, and improve overall team functioning. In aviation, these attributes are particularly salient during mixed-nationality operations, fast-changing environments, and ambiguous decision phases. Training programs can enhance this dimension through structured interventions:

- Scenario-based briefing simulations with intercultural complexity
- Reflective debriefing templates embedded in CRM cycles
- Peer-led journaling or experience sharing tied to known cultural variables
- Standardized MCQ assessments within CBTA or EBT systems

By cultivating metacognitive CQ across roles – flight crew, cabin crew, operations supervisors – aviation organizations reinforce adaptive learning. This does not replace procedural standardization; rather, it enhances contextual acuity within existing regulatory frameworks.

4.2.2 Cognitive CQ: Knowledge and Understanding

Accurate interpretation precedes effective adaptation. In cross-cultural settings, particularly in aviation, insufficient cultural knowledge can lead to miscommunication, coordination failures, and performance degradation. Cognitive CQ refers to the depth and structure of one's knowledge about cultural values, norms, communication patterns, and operational expectations (Ang & Van Dyne, 2008). This knowledge enables aviation professionals to form accurate expectations, adjust verbal and nonverbal strategies, and respond with contextual appropriateness under pressure. The relevance of this domain extends beyond polite interaction. Pilots and cabin crew must routinely interpret subtle cultural signals – how deference is expressed, when silence indicates agreement or dissent, or how decision-making unfolds in hierarchical systems. The absence of such knowledge can generate tension or delay critical feedback during time-sensitive operations.

Cognitive CQ (CCQ) encompasses both general cultural frameworks and domain-specific knowledge. General knowledge includes recognition of broad patterns (e.g., direct versus indirect communication, individualism versus collectivism, or attitudes toward uncertainty). Domain-specific knowledge, on the other hand, reflects localized familiarity: understanding the briefing norms at a particular base, expected forms of address with regional staff, or the preferred authority structures in a given airline's culture.

In practice, cognitive CQ enables crew members to:

- Anticipate differing interpretations of feedback, authority, or ambiguity
- Adjust tone, structure, or sequence of information to suit audience expectations
- Recognize situational norms around time management, task initiation, and information sharing
- Align with localized operational protocols, customs, and behavioral expectations

Professionals high in cognitive CQ are typically less reactive and more deliberate in their interpersonal engagements. They interpret unexpected behavior through an informed lens rather than defaulting to judgment or stereotyping. This enables smoother collaboration and improves CRM effectiveness.

Misapplication of cultural knowledge or reliance on heuristics undermines performance. The CCQ5 marker highlights patterns that should trigger corrective feedback in training or debriefing sessions. Even when technical operations proceed flawlessly, relational breakdowns can impair performance. A culturally unaware interpretation of respectful silence as disengagement, or of indirect phrasing as indecision, may result in missed escalation, eroded cohesion, or faulty assumption-making. To address this, CCQ should be embedded in Crew Resource Management CRM and EBT programs through:

- Case-based analysis of cultural misalignment in aviation incidents
- Role-play exercises simulating intercultural communication under operational constraints
- Visual decoding tasks, involving the interpretation of nonverbal cues across cultural contexts
- Briefing rehearsals that vary based on crew composition and destination-specific expectations

Training designed in this manner does not aim for cultural fluency in all contexts but seeks to foster agile interpretation and error-resilient communication. Over time, this contributes to operational synergy, particularly in multi-crew, cross-national configurations.

4.2.3 Motivational CQ: Interest and Drive

Cultural competence is not solely derived from knowledge or strategy; instead, it originates from a motivated approach to understanding. Motivational CQ serves as the impetus that propels individuals to engage across cultures with vigor, curiosity, and self-assurance. In the field of aviation, where teams are comprised of diverse individuals and time constraints are prevalent, this intrinsic motivation frequently influences whether an individual exerts the effort to connect or chooses to withdraw silently.

Motivational CQ encompasses the interest, perseverance, and optimistic mindset essential for adaptation and success in culturally intricate situations (Ang & Van Dyne, 2008). For crew members, possessing an understanding of cultural theory is insufficient; they must possess a desire to apply this knowledge, even in instances of stress or uncertainty.

The Motivational Cultural Quotient (CQ) refers to the intrinsic drive to adapt across cultural boundaries, specifically the degree of interest and confidence individuals exhibit in intercultural situations. It encompasses the following components:

- Intrinsic motivation: a sincere curiosity regarding other cultures.
- Extrinsic motivation: the aspiration to excel within international teams or in diverse work environments.
- Self-efficacy: the assurance in one's capacity to adjust and succeed in cross-cultural contexts

(Bandura, 1997).

In the field of aviation, motivational cultural intelligence (CQ) exerts a nuanced yet formidable influence. It significantly shapes the way crew members navigate differences, address misunderstandings, and dedicate themselves to teamwork. High levels of motivation facilitate proactive engagement, whereas diminished motivation may lead to withdrawal, frustration, or passive compliance.

Professionals exhibiting robust motivational CQ are more inclined to:

- Take the initiative to comprehend the backgrounds
- Demonstrate perseverance in the face of communication challenges
- Recognize diversity as an asset rather than a liability.

MCQ serves as the critical link between knowledge and behavior. In the absence of a genuine desire to connect and adapt, even the most effective CRM training may prove inadequate. In dynamic environments such as flight decks or cabin crew coordination, one's attitude frequently manifests with greater visibility than knowledge.

Consequently, cultivating motivational Cultural Intelligence should be a deliberate component of training and leadership development. Aviation training programs can enhance motivation by:

- Establishing psychologically safe environments for cultural exploration
- Acknowledging and rewarding cross-cultural initiatives
- Integrating personal storytelling and reflection into training procedures
- Providing mentorship and coaching focused on intercultural experiences.

Leaders can exemplify motivational cultural intelligence by exhibiting humility, posing inquiries, and demonstrating genuine curiosity. Such behaviors cascade down, fostering an environment in which others are encouraged to embrace, rather than fear, cultural differences.

4.2.4 Behavioral CQ: Action and Adaptability

Comprehending cultural norms is imperative; however, applying that understanding in real-life situations is what truly differentiates effective interactions. Behavioral CQ (BCQ) represents the outward manifestation of cultural competence: the capability to adapt verbal and nonverbal behavior to meet the requirements of various cultural interactions (Ang et al., 2007). In the field of aviation, where crews consistently operate under time constraints, BCQ has a significant influence on team cohesion, clarity, and, ultimately, safety.

This particular aspect of CQ illustrates not only individuals' knowledge or perceptions regarding cultural differences but also their behavioral responses to such differences. It delineates the distinction between having good intentions and being effectively understood.

Behavioral Cultural Intelligence (CQ) refers to the ability to interact effectively in culturally diverse environments. This competence involves adjusting one's speech, tone, body language, gestures, and overall communication style to align with the cultural norms of others (Livermore, 2011).

In the field of aviation, a high level of Behavioral CQ empowers professionals to:

- Provide clear and respectful instructions, regardless of the recipient's background.
- De-escalate tension by employing a culturally appropriate tone and posture.
- Cultivate trust by demonstrating awareness of local expectations and etiquette.
- Avoid behaviors that may be perceived as offensive or dismissive.

Behavioral CQ transforms theoretical awareness into real-time adaptation. It represents the juncture at which intercultural effectiveness becomes evident.

Behavioral Cultural Intelligence (CQ) is particularly paramount during high-stakes, time-sensitive operations, such as emergency briefings or in-flight conflict resolution. Crews that possess the ability to modify their communication styles to align with their colleagues' cultural expectations are significantly more likely to maintain composure, avert misinterpretations, and reassure passengers.

Flight crews, cabin personnel, and ground operations all benefit from behavioral agility. For instance, a ground controller who adjusts their phrasing for a pilot whose native language is not English can help prevent confusion. Likewise, a team leader who is aware of cultural preferences in nonverbal communication may effectively resolve crew friction while mitigating any escalation of tensions.

Crew Resource Management (CRM) training programs are increasingly incorporating simulation exercises, video reviews, and cultural behavior drills to bolster these capabilities.

4.3 INTEGRATING CQ DIMENSIONS IN AVIATION CONTEXT

Anastasios Plioutsias, Konstantinos Pechlivanis and Dimitrios Ziakkas

CQ does not function as four isolated traits. In aviation, it must operate as a system – an interplay of awareness, knowledge, drive, and behavior. Each dimension reinforces the others. Only when this integration occurs can crews adapt with precision, communicate without friction, and maintain safety across cultural boundaries. The industry's operational context demands more than individual competence. It requires crews to synchronize in real time. CQ provides the scaffolding for this synchronization. Metacognitive CQ helps crews detect emerging misalignments. Cognitive CQ prevents them from misreading intent. Motivational CQ supplies the energy to persist in complex interpersonal settings. Behavioral CQ makes the necessary adaptations visible and effective. Together, these dimensions create an environment where trust can form rapidly – even between strangers from vastly different cultural systems.

4.3.1 Practical Case Vignettes by CQ Dimension

The following vignettes illustrate how CQ dimensions manifest in aviation practice:

- **Metacognitive CQ – Reflective Briefing**
  A captain scheduled to lead a multinational crew reviews cultural profiles before departure. He slows his verbal pacing during the briefing and explicitly requests open feedback. Post-flight, he modifies his communication strategy based on observed reactions.
  Outcome: Increased participation during inflight phases; earlier recognition of procedural ambiguities.
- **Cognitive CQ – Awareness of Hierarchy Norms**
  During a turbulence diversion, a co-pilot from a high power-distance background withholds input. The captain, attuned to this dynamic, pauses decision-making to solicit feedback. The input reveals a safer alternate route.
  Outcome: Reduced operational risk; enhanced coordination under time pressure.
- **Motivational CQ – Feedback Engagement Across Cultures**
  A senior flight attendant initiates feedback from a new crew member from a distinct cultural background regarding procedural clarity. The dialogue reveals overlooked ambiguity in the initial instructions.
  Outcome: Later in the flight, the junior member independently identifies and reports a medical concern, improving passenger safety.
- **Behavioral CQ – Adapting During Evacuation**
  During an evacuation, confusion arises among passengers unfamiliar with the crew's gestures. The lead crew member adapts signals and simplifies verbal instructions.
  Outcome: Compliance increases; exit flow stabilizes without escalation.

These aren't abstract models. They're grounded responses – fast, precise, situationally aligned. What enables them is not technical skill alone but cognitive flexibility rooted in CQ.

These examples are not isolated demonstrations of competence. They represent the interaction of CQ dimensions applied in an operational context. Individuals who routinely plan, monitor, adjust, and learn from intercultural interactions – while sustaining motivation and behavioral adaptability – improve team dynamics and performance stability. The key observation is that CQ does not function as a modular checklist. It is a composite capability, inseparable from effective CRM execution, leadership, and interpersonal coordination. Aviation professionals who internalize this system are better equipped to interpret ambiguity, resolve latent cultural conflict, and maintain safety margins across variable operational conditions.

CQ integration is not a soft-skill aspiration. It is a necessity in environments where procedures are standardized, but people are not. The cockpit is not culturally neutral. Neither is the cabin, the ramp, or the tower. For these reasons, integration of CQ must extend beyond isolated modules. It should inform instructional design, leadership development, and recurrent training in accordance with CBTA and EBT standards. Where safety depends on mutual understanding, CQ integration becomes operationally indispensable.

To prepare crews for this complexity, training must mirror reality. CRM modules should build in cultural asymmetries. Briefings should include a structured debriefing on intercultural friction. Simulator scenarios must place interaction and interpretation on equal footing with checklists. Integrated CQ supports more than harmony. It underwrites precision. It reduces drift between intention and reception. It flags where hierarchy may silence input, where assumptions may mask risk, and where communication may stall under pressure.

Aviation safety rests on clarity. CQ delivers it – not abstractly, but through concrete, testable, teachable behaviors. The future of performance in culturally diverse airspaces depends not just on flying the aircraft but on reading the room. The essential conclusion is that CQ is not merely a checklist; it represents a mindset and practice that ought to be ingrained in every facet of aviation culture, encompassing CRM training, leadership development, and in-flight teamwork.

4.3.2 Implementation of the Theory in the Aviation Ecosystem

Appendices 1–3 present the necessary insights to practitioners for the integration of the proposed theory in the global aviation ecosystem. The tables provided interconnect the proposed theory with the Evidence-Based Training (EBT) and Competency-Based Training Assessment (CBTA) concepts (ICAO, IATA, EBT, CBTA).

4.4 CONCLUSION

Dimitrios Ziakkas

In the global aviation ecosystem, where cultural complexity is not an exception but a daily operational norm, the ability to navigate human difference is fundamental to safety, cohesion, and performance. This chapter aimed to deepen our understanding of CQ by detailing its four interconnected dimensions and examining their application within the context of aviation. We have further expanded the conversation by introducing DQ, recognizing that effective interaction cannot be limited to cultural knowledge alone. Crew members today face the complex demands of responding to generational, gender-based, cognitive, and socio-economic diversity – demands that require an evolved, more nuanced form of awareness.

What emerges from this exploration is not a set of isolated competencies, but a dynamic framework that enables professionals to anticipate, reflect, adapt, and lead in complex, high-stakes environments. When crew members engage in deliberate planning before interacting with others, when they monitor their behavior in real time, when they demonstrate persistence in seeking to understand, and when they adjust their verbal and nonverbal expressions accordingly, they are not merely following best practices. They are cultivating trust, reducing miscommunication, and enabling safer, more human-centered operations.

Importantly, CQ and DQ are not innate traits but trainable capacities. Their development must be embedded across the full arc of aviation training and professional development – from scenario-based CRM exercises to inclusive leadership education. Regulatory frameworks such as ICAO's CBTA and IATA's EBT provide fertile ground for aligning these human-centric competencies with observable pilot behavior and safety-critical outcomes. But such alignment will only be effective when supported by a cultural shift within organizations: one that values diversity not as an obligation, but as an operational strength.

As the aviation industry continues to evolve within an increasingly interconnected world, so too must our conceptual tools. The integration of CQ and DQ represents more than a theoretical advancement; it is a pragmatic response to the human challenges of multicultural collaboration in the cockpit, cabin, and beyond. It calls upon all aviation professionals – not simply to tolerate difference – but to engage with it actively, reflectively, and with purpose.

REFERENCES

Ang, S., & Van Dyne, L. (2008). Conceptualization of cultural intelligence: Definition, distinctiveness, and nomological network. In S. Ang & L. Van Dyne (Eds.), *Handbook of cultural intelligence* (pp. 3–15). Routledge.

Ang, S., Van Dyne, L., Koh, C., Ng, K. Y., Templer, K. J., Tay, C., & Chandrasekar, N. A. (2007). Cultural intelligence: Its measurement and effects on cultural judgment and decision making, cultural adaptation and task performance. *Management and Organization Review*, 3(3), 335–371.

Bandura, A. (1997). *Self-efficacy: The exercise of control*. Freeman.

Hughes, C. (2023). *Diversity intelligence: Leading with inclusive insight in a complex world*. Leadership Press.

Livermore, D. A. (2011). *The cultural intelligence difference: Master the one skill you can't do without in today's global economy*. AMACOM.

Livermore, D. A. (2024). *Leading with cultural intelligence* (3rd ed.). HarperCollins Leadership.

Thomas, D. C., Elron, E., Stahl, G. K., Ekelund, B. Z., Ravlin, E. C., Cerdin, J. L., … & Lazarova, M. B. (2008). Cultural intelligence: Domain and assessment. *International Journal of Cross Cultural Management*, 8(2), 123–143.

5 Cultural Differences in CRM and Communication

5.1 INTRODUCTION

Dimitrios Ziakkas and Alex Pollitt

There is a bewildering array of theories on intercultural communication. These are rooted in a fundamental controversy of how to define the concept of culture itself. Theorists from opposing schools of thought disagree on the fundamentals of the idea of culture, and this creates a problem in turn for defining what is meant by intercultural communication (Gudykunst and Mody, 2002). Two fundamental questions determine the different theoretical conceptualizations of culture, and consequently, how questions of culture are researched and investigated (Straub et al., 2002). The first of these is whether the structure of culture can be understood as a social phenomenon that exists separately from the individuals that make it up, as a coherent social unit, or whether culture is better understood as the collective product of the characteristics of individuals that form a cultural group. The second question is whether culture can be effectively explained in terms of the universal dimensions of a group, or whether it can only be properly described in terms of the characteristics of the individual, each of whom is influenced by a myriad of cultural strands (Schiffrin et al., 2001).

The Dutch academic Geert Hofstede called it "The collective programming of the mind which distinguishes the members of one group or category of people from another" (Hofstede et al., 2010). This definition reflects the views of the essentialist school of thought on the nature of culture. The essentialist view represents the traditional approach to understanding culture, based on an idea of culture as a tangible, concrete social phenomenon (Holliday, 2001). This school is also known as the social science approach because of its quantitative research methods and empiricist philosophy, and most research into intercultural communication, particularly in the more applied fields such as aviation, has been conducted within this theoretical framework (Holliday, 2001). Practical applications for Hofstede's cultural dimensions theory sparked a new sector of cultural awareness training and education grew up around the literature. The key message of this culture-generic training was that despite widespread differences between cultures, they differ in predictable ways. Hofstede's cultural dimensions, and other related models, have shaped the basic structure of the cross-cultural studies field and its controversies ever since (Peterson, 2004).

An opposing school of thought is made up of non-essentialist theorists, who reject the idea of presupposed cultural units as overly simplistic because they fail to account

DOI: 10.1201/9781003672029-7

for the complex and subjective nature of human behavior. This group argues that although essentialist definitions are satisfactory for explaining behavior at a macro level, they do not adequately describe behavior at the individual level (Straub et al., 2002). A fuller understanding of culture must acknowledge the affluence, complexity, and intertwinedness of variables and contexts that determine how a person's character is manifested.

The opposing theoretical perspectives on how to define and understand culture reveal an inevitable trade-off between achieving a true and encompassing definition of culture on the one hand and achieving useful theoretical paradigms capable of wielding explanatory power on the other. Advocates of the social science approach (Hoecklin, 1995; Hofstede, 2003; Trompenaars & Hampden-Turner, 2020) are generally interested in explanation. A certain level of generalization and assumption is required to allow for the creation of cultural groups. Groups can then be tested for universal traits and values from which models can be derived and conclusions drawn about differences or commonalities to facilitate understanding. This is both what allows us to apply cultural knowledge to practical scenarios and the main source of criticism for this approach, provoking accusations of stereotyping and presupposition (Baumann, 1996). The difficulty of a more all-encompassing theory of culture is that it loses its explanatory power.

In aviation, definitions of culture have focused almost exclusively on essentialist explanations of the concept, with most research grounded in Hofstede's model (Pratama & Caponecchia, 2025). Aviation, grounded in the practical, is unlikely to have sympathy with more complex and perhaps abstract theories of culture. Its interest in cultural communication is a practical one. Nevertheless, the student of culture in aviation should establish at least a working knowledge of other interpretations and have a grasp of the sheer complexity of the concept of culture and the different perspectives from which it is viewed.

Chapter 2 introduced the theoretical foundations of how culture is understood in the context of aviation. A social scientific approach centered on Hofstede's cultural dimensions dominates both how culture has been studied and how it is taught to flight crews. It is not the aim of this chapter to go into greater depth about types of cultural dimensions and the place of different nations on relative cultural scales. Instead, it focuses on what we know from research and operations about national cultures in the aviation domain, and more specifically, how culture interacts with the work of pilots on the flight deck. What do we know about national cultural differences operationally, and what does this tell us about how to maximize operational efficiency and safety in the ever more multicultural environment on board our aircraft? This chapter explores the complex and evolving dynamics of multinational crew operations, with a particular focus on how cultural diversity influences crew performance, coordination, and safety outcomes. Grounded in the principles of CRM, CBTA, and EBT, the discussion is framed around both theoretical perspectives and operational realities. A historical lens is used to trace the development of diverse crew compositions globally, with an emphasis on how these dynamics have unfolded in the Middle East – a region that hosts some of the world's most multicultural cockpit environments.

The analysis moves beyond surface-level observations, engaging deeply with cultural theories, human factors, and regulatory frameworks from ICAO, IATA, EASA, the FAA, and the UK CAA. It also considers the implications of emerging technologies, including artificial intelligence and human–machine interfaces, in shaping the future of multinational crew operations. By examining these factors within a coherent human-centric framework, this chapter aims to offer both critical reflection and practical insight into how cultural intelligence can be leveraged as a safety resource in global aviation.

5.2 MULTINATIONAL CREW DYNAMICS

Alex Pollitt and Dimitrios Ziakkas

There are three types relevant to aviation culture: national, organizational, and professional (Helmreich and Merritt, 2017). For the purposes of this subchapter, however, this discussion is limited to national differences. National culture refers to how a person's geographic country of origin shapes their attitudes, behaviors, and social conduct. In its early days, aviation was a largely nationalistic endeavor. National carriers had primarily national crews. The introduction of jet airliners made long-haul travel more accessible, and in the following decades, the emergence of truly international airlines marked a shift toward the global system we know today. Some of these airlines now boast more than 70 nationalities among pilots and crew (Cathay Pacific, 2025), and multinational flight crews have become a hallmark of global aviation.

Aviation's interest in culture is a practical one and has been driven principally by risk management and the pursuit of safety. This utilitarian approach requires methods for identifying cultural differences, understanding their implications, and teaching cultural competence based upon them. However, as Part I has already explored, culture is a topic of remarkable complexity, both in breadth and depth, and consequently most explanations of culture are bound to help people make sense of this complexity.

Everyone is shaped by the norms and behaviors of the culture where they grew up, and the impact of this can be thought of in terms of two effects. When brought on to the flight deck, the effect of social/interpersonal attributes could include failure to communicate information effectively or contribute to disagreements or conflicts in decision-making. Cognitive effects, on the other hand, might include the misinterpretation of a critical cue, mis-assessment of a situation, or the implementation of an erroneous response (Strauch, 2010). The impact of linguistic diversity could be identified as a third factor, and this will be discussed in more detail later in Section 5.3.

Cultural factors have contributed to aviation accidents over the years and are frequently implicated in safety outcomes (Strauch, 2010; Li et al., 2009; Merrit & Helmreich, 1996). As the multi-national composition of the global aviation system has continued to evolve and the industry becomes ever more integrated, the influence of culture in flight safety will not diminish. Perhaps unsurprisingly, therefore, interest in the impact of national culture on the flight deck has also grown in recent years and is now the focus of a large body of research on culture's consequences, much of it focused on national differences. What practical insights do all these studies of culture on the flight deck give us?

5.2.1 Historical Emergence of Diverse Flight Crews

The journey toward multinational crew configurations has been incremental and uneven, shaped by geopolitical shifts, economic liberalization, and evolving labor markets. In the post-World War II era, airlines such as Pan American World Airways and British Overseas Airways Corporation (BOAC) symbolized national identity, employing almost exclusively home-country nationals. This homogeneity reinforced shared professional cultures but limited the adaptability and scalability of airline operations across international markets. By the late 20th century, deregulation (see the Airline Deregulation Act in the United States 1978) facilitated the entry of new carriers and altered traditional employment patterns. The subsequent rise of global alliances (e.g., Star Alliance, oneworld) and transnational ownership models (e.g., LATAM, IAG) compelled operators to recruit internationally to support round-the-clock global operations. Middle East carriers (e.g., Emirates, Etihad, and Qatar Airways) exemplified this shift.

5.2.2 The Middle East Case Study

The Gulf Cooperation Council (GCC) states have emerged as aviation powerhouses over the past three decades. Their strategic geographic position as global connectors, combined with state-backed airline growth and modern infrastructure, attracted international aviation talent. However, the regulatory frameworks governing flight operations in these jurisdictions often rely on hybrid models (i.e., EASA-aligned standards merged with regionally tailored practices). Regional variance in power distance is particularly salient in Middle Eastern airlines, where crews from Asia, Europe, and Africa operate under a unified SOP but with vastly different cultural interpretations of assertiveness.

5.2.3 Regulatory Frameworks

Global aviation regulators have increasingly acknowledged the influence of multicultural crew dynamics on operational safety. The International Civil Aviation Organization (ICAO), in Doc 9683 "Human Factors Training Manual," and Doc 9995 "Manual of Evidence-based Training," encourages member states to integrate cultural sensitivity and behavioral competencies into flight crew training (ICAO, 2013). These documents underpin a global philosophy that, while standard operating procedures (SOPs) ensure a minimum level of uniformity, human performance varies by cultural context and must be acknowledged explicitly in training design.

The European Union Aviation Safety Agency (EASA), through regulations such as AMC1 ORO.FC.115 and AMC1 ORO.FC.220, embeds CRM and CBTA principles in pilot training and assessment. Operators are required to incorporate behaviorally anchored markers aligned with observable competencies. This allows for the detection and remediation of culturally rooted communication or decision-making gaps (EASA, 2020). The UK Civil Aviation Authority (CAA), although historically aligned with EASA, has maintained an independent human factors oversight function and supports the integration of cultural performance factors into simulator assessment criteria.

In the United States, the Federal Aviation Administration (FAA) mandates CRM training for Part 121 carriers under Advisory Circular 120-51E. While not explicitly cultural in its orientation, FAA guidance emphasizes assertiveness, shared mental models, and error management – all of which have been operationalized in training programs to counteract deference-based miscommunication, particularly in international crew pairings.

IATA complements these efforts with its "Guidance Material and Best Practices for CBTA Implementation," which advocates for the adaptation of training scenarios to local cultural norms while retaining a global competency benchmark. This balance allows multinational operators, particularly in culturally complex environments like the Middle East or Asia-Pacific, to reconcile standardization with adaptability.

5.2.4 The Cultural Biases: What The Research Tells Us

Attitudes and interactions on the flight deck are intertwined with national culture which directly (and indirectly) influences safety and risk-related behaviors such as effective communication and decision-making. Accordingly, the focus of much research has been on the negative impact of culture and the challenges it presents to safety. A recent review of scientific literature from the last 20 years exploring the role of national cultures in aviation safety (Pratama & Caponecchia, 2025) found that 30% of studies aimed to establish a link between national culture and error, incident, or accident statistics, and nearly two-thirds were about the characteristics of national culture and how they affect flight crew performance, communication, teamwork, and safety culture.

It is noteworthy that almost all the studies reviewed (97%) used a framework for measuring cultural differences based on selected elements of Hofstede's model of cultural dimensions (Pratama & Caponecchia, 2025). Research has repeatedly shown three relevant dimensions. These were Power Distance (which measures an individual's relationship with the exercise and acceptance of authority and hierarchy); Individualism vs. Collectivism (which describes a person's tendency toward individual or group behaviors); and Uncertainty Avoidance (which measures how comfortable a person is with unknown, unpredictable, or ambiguous situations). Such is the dominance of Hofstede's dimensions of national culture in the methodology of studies on culture in aviation that one or more of these three dimensions were the focus of over three-quarters of research (89% examined Power Distance, 77% Individualism–Collectivism, and 71% Uncertainty Avoidance).

5.2.5 Challenges of Diverse Flight Crews

Safety performance still varies between countries (ICAO, 2021), and national culture is one of the variables affecting those safety outcomes. A disproportionate amount of research has focused on just the three aforementioned cultural dimensions because they are considered contributors to team performance (Klein, 2005). Team performance suffers the greatest in phases of high operational workload (Urban et al., 1996). Under pressure, when social conscience is less able to repress cultural norms, we tend to "revert to type."

Power Distance, Individualism, and Uncertainty Avoidance have all been found to impact both team communication and decision-making processes. Multiple studies on Power Distance have shown it to affect cockpit gradient through crews' perceptions of hierarchical status and the way subordinates view and defer to members of their team with greater authority (Reader et al., 2015). Collectivism impacts team dynamics, expectations of leadership styles, and decision-making processes. For example, a high collectivism culture which esteems group harmony could result in a failure to challenge unsafe activity through a reluctance to "rock the boat." A more collegiate approach to decision-making has also been reported to have a negative effect on outcomes (Hodgson et al., 2013). Put a high degree of Power Distance and a high degree of Collectivism together and multiple studies found there was a significant correlation with a higher accident rate (Al-Wardi, 2017, Chan & Li, 2021, Enomoto & Geisler, 2017, Jing et al., 2001), a finding consistent with studies outside of aviation as well, including road traffic and maritime accidents (Atchley et al., 2014; Kwon, 2019).

Uncertainty Avoidance is significant because socio-technical systems, by their nature, generate complex, uncertain, and ambiguous problems. Where a person sits on the Uncertainty Avoidance scale determines how they manage ambiguity both cognitively and socially. For example, cultures with high uncertainty avoidance are more dependent upon structured rules and standard operating procedures and are less confident operating outside of these frameworks with creativity or flexibility should the need arise. This creates elevated risk in the case of non-normal or emergency scenarios; hence, Uncertainty Avoidance has been associated with higher incident rates (Soeters and Boer, 2000; Stancu, 2013) and greater levels of organizational/supervisory error (Al-Wardi, 2017).

5.2.6 The Importance of Cultural Relativity

Popular awareness of culture's impact on aviation safety is dominated by the Power Distance dynamic. A captain from a hierarchical society carries cultural baggage: the expectation of a proper respect for authority. He does not expect to be questioned, and his co-pilot, perhaps enfeebled by his own cultural inheritance, does not imagine doing so. While not incorrect, this view of high Power Distance and its negative association with CRM is an oversimplification. Power Distance has consistently been shown to negatively impact desirable behaviors, ranging from CRM (Peksatici, 2018, Seva et al., 2007), crew communication (Hazrati, 2015, Noort et al., 2021; Ritsher, 2005; Zhu and Ma, 2015; Hodgson et al., 2013), team performance (Helmreich, 2000), and safety culture and climate (Ben-Saed & Pilbeam, 2022; Liao, 2015; Reader et al., 2015; Tear et al., 2020). However, this creates a tendency for members of low Power Distance countries to look at these dynamics as someone else's problem, and to pat themselves on the back for their own low Power Distance cultural inheritance.

Most cultural dynamics are viewed (and have been studied) through a Western cultural lens (Jing & Batteau, 2015). Consequently, Western cultures tend to think of the dimensions discussed above as being particularly associated with Eastern cultures. This is not the case. Understanding that every country takes its place on the scale of each dimension is critical to interpreting the nature of the interaction between colleagues and co-workers. For example, Power Distance can still represent an obstacle to effective communication for two European pilots together on a flight deck, provided their relative position on the scale is separated. A Scandinavian pilot is likely to have a very different understanding of good hierarchical relationships than a Polish colleague because, in Scandinavia, organizational structures are relatively flat and good leaders are seen to be facilitators amongst peers. In Poland, on the other hand, the ideal relationship between leader and subordinate is probably a relatively strong authority figure who can give firm direction, and organizational structures are expected to be more hierarchical and fixed.

The same is, of course, true of other cultural dimensions. When it comes to communicating, in some countries, good communication is clear and precise, and messages should be understood at face value. They require little accounting for nuance or context in the communication style. The Netherlands is an example of a low-context country. When imparting or receiving a message, the straight-talking Dutch expect that it to be frank, direct, and to the point, with little need to dress or moderate language to account for the sensibilities of an interlocutor. Their close neighbors, the French, however, are a long way down the relative scale of communicating toward the high context end. In France, good communication is nuanced and layered. Messages are often implied rather than expressed plainly, and participants are expected to read between the lines.

Debriefing is a critical skill in aviation for both learning and safety, and is practiced daily in most operations, particularly during training and checking cycles. How we evaluate and give feedback is also fraught with cultural difficulty. Different cultures accept and expect different levels of direct feedback. On one side of this scale are countries such as Russia and the Netherlands where negative feedback is provided frankly, bluntly, and honestly, with no expectation to soften these messages with positive ones. In these cultures, it is also more acceptable to give criticism openly in front of a group. At the other end of the scale, in national cultures such as Japan or Indonesia, such an approach would be entirely inappropriate and considered offensive. In these cultures, a more diplomatic approach is expected, with negative feedback wrapped in positive messages and softer language. Here, even professionally constructive criticism is never given in public (Meyer, 2016).

Cultural relativity is probably the most important thing to understand about successful cross-cultural communication in an applied scenario, such as an aircraft crew working together, and what matters is the relative position of cultures along a scale, not absolute position. Furthermore, in such a highly international context as aviation, in the real world, intercultural communication does not take place bilaterally between two nationalities on a binary scale. A crew might consist of multiple different nationalities and cultural standpoints which need to be accounted for.

5.2.7 Benefits of Diverse Flight Crews

It is interesting to observe that in other work domains outside of aviation, the research exploring multinational team diversity, which has not focused on the negative impacts on performance, emphasizes a significantly more positive contribution. In a less specific context, team diversity has been shown to offer increased creativity and better problem-solving capabilities (Cox & Blake, 1991). Multiple studies have provided evidence for this "Diversity Bonus" (Page, 2017) which occurs when a group of individuals endowed with a range of cognitive approaches is more likely to bring a unique toolset to resolving a problem or situation. Evidence shows there is a higher possibility of a diversity bonus when a task is complex and requires deeper technical and cross-disciplinary knowledge. Helmreich and Merritt (2017) also maintained that well-managed cultural diversity leads to more thorough cross-checking, prevents complacency, and reduces breakdowns in CRM.

Additionally, Seva et al. (2007) found that Collectivism and Uncertainty Avoidance had a positive influence on crew resource management, as a strong group ethos reinforced CRM's message about teamwork and crew coordination. It has also been suggested that Uncertainty Avoidance is associated with a lower rate of crew errors because conformity to rules can suppress errors generated by deviance (Pratama et al., 2018) and support aspects of crew resource management. However, knowing how to access these benefits is fundamental to high-performing teams. In the early stages of bringing together team members from different cultural backgrounds, there is a tendency for people to default to in-group/out-group dynamics based on nationality, language, or culture. To overcome this effect, time is needed as a buffer against any initial friction caused by cultural differences. Helmreich et al. (2010) described a "Familiarisation Effect" where time spent together allows culturally diverse crew members to build trust, learn each other's communication patterns, and develop shared mental models. A functional understanding of interpersonal strengths and limitations has been shown to boost safety and efficiency.

5.2.8 EBT and CBTA: Competency Meets Culture

EBT and CBTA reflect a paradigmatic shift in pilot development – from traditional hour-based metrics to demonstrable competencies grounded in operational relevance. This evolution enables instructors to address not only technical proficiency, but also behavioral tendencies influenced by culture. For example, within CBTA, the competency "Leadership and Teamwork" can be assessed through behaviors such as "encourages input from all crew members" or "maintains open communication regardless of rank." These markers provide a framework for identifying whether cultural reluctance to challenge authority is interfering with flight safety.

EBT, as practiced by Lufthansa and Qatar Airways, among others, utilizes recurrent training data to tailor scenario-based exercises. These scenarios are built around actual operational risks and behavioral trends, making training not only data-driven but culturally attuned.

As ICAO and IATA continue to promote EBT and CBTA globally, the need for instructor standardization becomes critical. Not all instructors are equipped to recognize cultural behaviors as distinct from personality traits or competence issues. Thus, instructor development programs must include modules on intercultural communication, bias recognition, and inclusive debriefing techniques (EASA, 2021).

5.2.9 Emerging Technologies, AI, and the Cultural Ergonomics of Crew Interaction

Artificial Intelligence (AI) systems are becoming increasingly integrated into aviation operations. New opportunities – and risks – arise concerning cultural interaction with technology. Digital co-pilots, decision-support tools, and wearable cognitive monitoring systems are poised to augment flight safety. However, these systems must be designed and deployed in ways that are culturally ergonomic, ensuring that their interpretations and interventions align with the behavioral expectations of diverse users.

For instance, voice-command systems trained on Western speech patterns may misinterpret tonal inflections common in Southeast Asian or Middle Eastern speech, resulting in misrecognition during high workload phases. Similarly, AI-generated advisories that appear to override human command could be perceived as intrusive or disrespectful in high power distance cultures, undermining trust and usability.

The concept of "cultural ergonomics" emphasizes the interface between culture, cognition, and system design. In aviation, this means cockpit displays, alert systems, and even AI-generated CRM feedback must be tested for cross-cultural interpretability. For example, Airbus' prototype Human–AI Collaborative Interface includes culturally adaptive dialogue management protocols, allowing the AI to modulate tone and formality based on crew profile.

Moreover, wearable systems that monitor fatigue or stress levels can be culturally sensitive in their feedback mechanisms. A biometric alert suggesting that a crew member is "underperforming" may be received differently across cultures – ranging from an opportunity for self-correction to a perceived insult. Therefore, the framing and delivery of such feedback must be adapted accordingly.

5.2.10 Toward a Multicultural Safety Culture

The multinational cockpit represents both a triumph of global cooperation and a crucible for cultural complexity. Diversity in aviation is not a peripheral consideration – it is central to operational integrity in an interconnected world. When flight crews comprise individuals from varied cultural backgrounds, their collective ability to manage ambiguity, distribute workload, and co-create safety is shaped by the degree to which training, regulation, and technology accommodate their human differences.

This chapter began by noting the almost uniquely global nature of the aviation industry. If we accept the premise that the need for cultural competence grows with the number of different nationalities making up a team, then aviation should perhaps be leading the world in this area of expertise! As an industry, it has been slow to pick up on the importance of understanding its extraordinarily multicultural DNA. For most people, the norm is still to view the world around them from the comfort of their own cultural perspective in the uncritical assumption that all misunderstandings and disagreements, alternative mentalities and methods, are rooted in personality and individual differences. However, the evidence tells us otherwise: culturally rooted core belief systems and patterns of behavior drive perception (what we see), cognition (what we think), and actions (what we do), all of which are taken into the air by flight crews.

It is impossible, without accounting for the myriad factors that create the context in which intercultural interaction takes place, to conclude that having diverse nationalities or cultures on a team is a net positive or negative to safety. Research from both inside and outside of aviation tells us that there are potential risks and potential benefits when combining different nationalities in teams on board aircraft. For aircrew, this raises two practical questions: 1) how to identify which of these might be at play in your team, and 2) how best to harness the benefits and minimize the risks. The task of developing cultural intelligence through training and operational experience is to raise both individual and team understanding of how to answer these two questions. If a team is to be effective in managing culturally introduced risk and maximizing the "diversity bonus," its members must first understand the cultural context that shapes their team dynamics and then consider how it impacts their operational efficiency and safety.

Culture is weaved into the fabric of all the social interactions taking place within the aviation's socio-technical system (Leveson, 2004), yet despite a multitude of studies on culture in aviation, we still understand little about how the cultural diversity influences the socio-technical system as a whole. Although cultural awareness is acknowledged as a desirable behavior within the pilot competencies (ICAO, 2013), it is not a flight crew competency of its own. It currently sits within the competency of Leadership and Teamwork (LTW), but its influence clearly impacts broadly across other competencies as well. As yet, no studies have attempted to identify a comprehensive set of behavioral markers describing how culture interacts with other established flight crew competencies such as Communication, Leadership and Teamwork, Problem-solving and decision-making (Pratama & Caponecchia, 2025). However, building cultural competency across the non-technical skills for both individuals and teams can be thought of as a three-part inverted triangle, starting with self:

1. Develop cultural-self-awareness: know yourself.
2. Develop awareness of the cultural inheritance of other members of your team.
3. Consider the impact of the balance of cultures within your team.

Emerging models such as CBTA and EBT, under the guidance of ICAO, IATA, EASA, and national authorities, offer a scaffold for embedding cultural intelligence into competency frameworks. However, their success hinges on the ability of organizations to invest in instructor development, data-driven scenario design, and open dialogues about culture and performance. In parallel, emerging technologies – particularly AI-driven systems – must be designed with cultural ergonomics in mind. These tools must not merely function across contexts; they must resonate with the people who use them. The risk of technological alienation or miscommunication is amplified in culturally diverse crews unless inclusivity is embedded at the design level.

Ultimately, safety is a collective accomplishment. It is built not only on procedures and checklists but on mutual understanding, adaptive communication, and respect for differences. As aviation continues to evolve, embracing the complexities of multicultural crew dynamics will not be optional – it will be essential.

5.3 LANGUAGE AND COMMUNICATION BARRIERS

Spiros Chazapis

Aviation demands consistent performance under conditions that often test psychological endurance. On the late afternoon of March 27, 1977, two fully loaded Boeing 747 aircraft, operated by Pan Am and KLM, respectively, were preparing for departure from Los Rodeos Airport on the Spanish island of Tenerife. It had been a long day for the passengers and crews of both aircraft – they were originally bound for the Gran Canaria Airport but were forced to divert to Tenerife due to a security incident. The weather at Los Rodeos was far from ideal. A low layer of fog and drifting clouds caused visibility to fluctuate significantly, at times dropping to as little as 100 meters. Both flight crews and the air traffic controller on duty were from their respective countries, with the Pan Am crew being the only native English speakers.

Following a long wait for refueling, passenger boarding, and other necessary procedures, the KLM crew finally received the communication they had been long waiting for – a clearance for departure. Shortly after, the Dutch aircraft began its takeoff roll. However, a critical misunderstanding had occurred. A combination of unclear communication, radio interference, ambiguous phraseology, and an overly steep command gradient in the flight deck misled the KLM crew into believing they received a clearance that had not, in reality, been issued. Despite last-minute evasive attempts by both crews (the PAN AM aircraft was still on the runway, having missed its previously assigned exit taxiway due to low visibility), the two aircraft collided. To this day, this accident remains the deadliest in aviation history.

Commercial aviation is a truly international industry. Every day, thousands of people from diverse cultural backgrounds have to work together in order to transport passengers and cargo across vast distances in a safe, efficient, and comfortable manner. Commercial aviation is an inherently multicultural environment. This *system of systems* is characterized by significant cultural and linguistic diversity, which, in most cases, can be considered a factor that helps reduce culture-induced systematic errors and, therefore, promote safety. Effective communication is a critical safety component. Miscommunication between pilots, air traffic controllers, and ground staff has been identified as a leading contributor to aviation incidents, including runway incursions and midair collisions. Conversely, clear and structured crew interaction has been shown to promote timely decision-making and coordination in complex airspace environments (Wickens et al., 1998) by fostering common mental models that support effective monitoring and intervention. It is, therefore, considered that standardized phraseology, adherence to communication protocols, proficiency in English, and clear articulation are indispensable to aviation safety.

As aviation continues to globalize, understanding and managing cultural differences remain essential to maintaining and improving safety standards (Kanki, 2019). This section examines the challenges of using verbal communication for operational purposes in a multicultural environment. The effects of inefficient verbal communication are discussed, along with mitigation strategies to minimize their impact on aviation safety.

5.3.1 Language and the Communicative Process

Human communication is a highly complex process involving the exchange of messages through both verbal and non-verbal means. In aviation, nevertheless, communication can only be conducted verbally, depriving the process of an important non-verbal component that humans are especially attuned to – potentially leading to miscommunications that have been implicated in several aviation incidents.

Language, in its broadest sense, is defined as "the systematic, conventional use of sounds, signs, or written symbols in a human society for communication and self-expression." Put differently, language represents a structured system of shared symbols used to transfer information, express thoughts, and convey emotions – each constituting an important aspect of human social existence. It is important to note that language is deeply embedded in the cultural context of its speakers, reflecting their values, norms, and thought patterns. According to the theory of high- and low-context cultures, even the very quantity of information transmitted through language varies between cultures – with individuals from high-context cultures relying heavily on implicit messages, shared knowledge, and non-verbal cues, which can lead to misunderstandings in communication with persons that have a cultural predisposition toward explicit, direct, and unambiguous exchanges. This means that the messages conveyed through language are not universally interpreted; rather, they are understood within the framework of cultural expectations, shared knowledge, and the overall worldview of the individuals engaged in the communication process. Moreover, effective communication is characterized as a linear process comprising four primary components: a sender, a message, a medium, and a receiver. Most contemporary models of communication also incorporate a feedback loop – considered an independent communicative act – to preserve the integrity of the interaction and facilitate mutual understanding. Each element within this process represents a potential point of failure; miscommunication may occur during the encoding or decoding phases (attributed to the sender and receiver, respectively), or during transmission, often due to constraints inherent in the communication channel. All these factors can have significant repercussions in the context of aeronautical communications.

5.3.2 English: The Language of Aviation

English language ability is often categorized into three different levels of proficiency. Individuals who have acquired English from early childhood as their first language are usually categorized as Native English Speakers (NES). English as a First Language

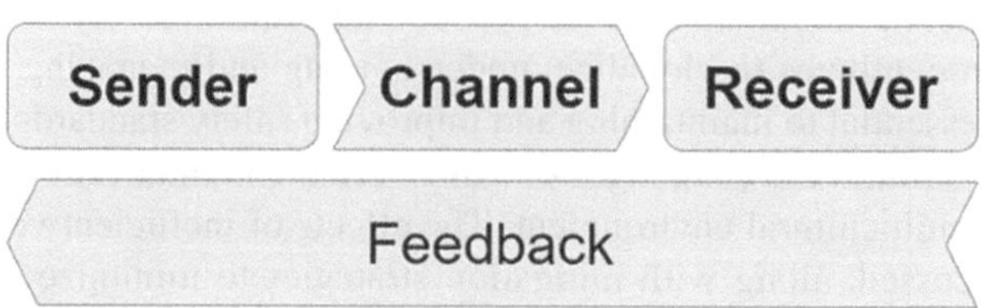

FIGURE 5.1 The Communication elements diagram.

(level EL1) is like NES and assigned to individuals who possess fluency equivalent to native speakers, usually the result of being raised in multilingual families or in countries where English is in widespread use in education or administration (for instance, India). Conversely, English as a Second Language (EL2) speakers are individuals who have learned English after acquiring a different first language. Introduction to English for these individuals usually occurs in late childhood or early adulthood as part of formal training in school or a language institute. These distinctions are important in aviation contexts, as differences in proficiency can affect clarity, comprehension, and the ability to handle non-routine or unexpected communication, especially under stress or in multicultural settings.

Establishing a global standard for language proficiency in aviation is not a trivial matter. Although English has evolved as the international working language of aviation, it is not the most widely spoken language globally – particularly at the native-speaker (NES/EL1) level. Furthermore, the number of speakers of English as a second language (EL2) now surpasses native speakers by a ratio of approximately 3:1. As a result, interactions between native and non-native speakers (NES/EL1–EL2), or between non-native speakers (EL2–EL2), are highly likely.

The International Civil Aviation Organization (ICAO), recognizing the critical role of effective communication in ensuring aviation safety, introduced language proficiency requirements early on in its history. Provisions were made as early as the Chicago Convention (1944), which, in Article 33, required proof of language proficiency for flight crews. ICAO further addressed language issues through two Recommended Practices in Annex 10 and a Standard in Annex 1, which focused on ensuring communication between aircraft and ground stations that did not share a common language. These provisions called for the availability of English as a common language and required that air traffic controllers be able to communicate in the nationally designated language(s) "without accent or impediment which would adversely affect radio communication." Notably, no explicit proficiency standards were required for flight crews at that time. Further work on formal English Language Proficiency (ELP) requirements began in 2000, leading to the implementation of first-generation ELP programs in 2008. Widespread compliance was achieved by 2013, although many states required extensions and technical assistance to fully implement the standards.

At present, personnel required to use radiotelephony are prohibited from engaging in operational duties unless their licenses are endorsed with an ICAO-recognized English proficiency level, ranging from Level 4 (operational) to Level 6 (expert). These endorsements are valid for a limited duration – except for Level 6, which does not require revalidation – and are subject to recurrent evaluation every 4–6 years, depending on the level achieved (ICAO, 2013). Virtually all ICAO contracting states have now adopted this requirement which remains a prerequisite for employment in the pilot or air traffic control roles.

5.3.3 Language as a Tool in Aviation Communication: English Versus Aviation English

While English functions as the default language of aviation – a *lingua franca*, or common means of communication between speakers of different native languages – aviation English constitutes a specialized variety distinguished by limited vocabulary, strictly defined phraseology, and constrained functionality. Marked differences exist between general English and aviation English, with the latter often deviating from standard grammatical rules to such a degree that untrained speakers may struggle to comprehend transmitted messages. These deviations are intentional, designed to reduce ambiguity, minimize misinterpretation, and promote clarity, predictability, and brevity in safety-critical operational environments. Additionally, limitations inherent to the communication medium – typically radio transmissions operating in VHF or HF frequency bands, – necessitate adaptations such as the phonetic alphabet (e.g., Alpha, Bravo) to enhance intelligibility over noisy or degraded channels. These constrained linguistic structures have proven effective in supporting communication across international airspace, where interlocutors often possess varying levels of English proficiency and diverse native language backgrounds. However, the rigidity of Aviation English also presents a key vulnerability: only a limited number of operational scenarios can be anticipated and codified into standard phraseology. Communication challenges emerge particularly in non-routine or emergency situations, where interactions extend beyond standardized expressions and require flexible language use – something that non-native speakers may find difficult to manage.

5.4 COMMUNICATING ACROSS CULTURES

Spiros Chazapis

This subchapter examines how the nature of air traffic control (ATC) communications and the linguistic variations have been identified as a causal or contributory factor in aviation accidents and incidents. Typical examples include collisions – both on the ground and in flight – controlled flight into terrain, and violations of clearances, such as unauthorized altitude deviations or incursions into restricted airspace. Gibson categorized communication errors into three top-level domains: communication error criteria, level of grammar, and external error modes. Particularly relevant to intercultural communication is the second one, which can be further subdivided into phonological errors (including prosodic features such as intonation, pausing, volume, and speech rate), syntactic errors (e.g., misapplication of morphology), semantic inaccuracies (patterns of language that affect meaning), and pragmatic failures (context-dependent use of language). At a more practical level, an analysis conducted by EUROCONTROL identifies several recurring contributors to pilot–controller miscommunication. Among the most common are accent, speech rate, use of non-standard phraseology or idiomatic expressions, and expectation bias. Similarly, Wever emphasizes that ambiguous phraseology, language difficulties, and communication between non-native interlocutors are significant sources of error. According to ICAO, efforts to standardize aviation phraseology may be insufficient to counteract deeply ingrained cultural and linguistic traits – such as accent, contextual interpretation, and language-specific idiosyncrasies – especially when semantic processing speed is affected by high-stress conditions, cognitive load, or fatigue.

5.4.1 Dialects, Pronunciation, and Regional Variations

Instituting a standardized language in aviation does not automatically eliminate the effects of variations in dialect, differences in pronunciation, intonation patterns, and regional idiosyncrasies. Diverse linguistic backgrounds pose a significant challenge to communication between crew members and air traffic controllers, indicating that communication efficiency decreases by nearly 23% when there is no shared linguistic background – particularly in high-workload or non-routine/emergency situations. This reduced efficiency and overall lack of precision can lead to outcomes ranging from minor inconveniences, such as requesting a transmission to be repeated over a congested frequency, to serious events like altitude deviations or runway incursions (Barshi & Farris, 2013). Their effect is further compounded by the inherent limitations of radiotelephony, a unidirectional medium that deprives users of the non-verbal component of communication and often suffers from poor audio quality. The following phonological and linguistic areas have been identified as critically important to mutual intelligibility in aviation:

- **Pronunciation and phonetic variations**
  This is one of the most significant, easily identifiable, and resilient aspects of verbal communication. Differences in pronunciation and the resulting communication difficulties are a staple of comedy, a situation anyone can easily relate to, regardless of cultural background. A preference for the sounds used in a person's native language emerges early in development, influencing the production of vowel and consonant sounds. Efforts, later in life, to mimic different accents or articulate unfamiliar sounds can be challenging – actors, for example, often undergo extensive training to master such skills. In some cases, a pronunciation that is close to native but perceptibly "off" may even result in an "uncanny valley" effect, a phenomenon usually associated with human-like robots or avatars. This can cause listener discomfort and hinder communication, as both the sender and receiver may struggle with encoding, decoding, and providing feedback on the transmitted messages. Phonetic variations occur in many forms, with some typical examples being:
  - **Vowel distinction problems**: Many users of aviation English struggle with vowels that do not exist in their native language. Typical examples include the sounds /ɪ/, as in pitch, versus /iː/ as in peach (affecting Spanish, Japanese, Arabic, and Korean speakers), the sounds /ɛ/ as in set versus /eɪ/ as in seat (for French and Turkish speakers), and /æ/ as in flaps versus /ʌ/ as in flops (Russian, Chinese, and Japanese speakers).
  - **Consonant and consonant clusters**: Similarly, typical examples include the inability of Arabic speakers to differentiate between the sounds /p/ as in pressure and /b/ as in bearing. Japanese speakers often have difficulty distinguishing between /l/ (as in left) and /r/ (as in right), as the language has a single liquid consonant, /r/, that lies somewhere between the two. Furthermore, several languages avoid consonant clusters (as in street or glide) altogether, and proceed to break them up by inserting vowels

(epenthesis) in between them – for instance, Japanese pilots often insert vowels between consonants ("cleared for takeoff" becoming "cu-leared for takeoffu"), affecting message comprehension.

- **Intonation patterns**: Languages differ significantly in sentence tone and stress. Many Asian languages, for instance, rely heavily on intonation patterns to convey meaning, making strict adherence to tonal rules essential for accurate understanding. In contrast, Germanic and Romance languages such as Spanish and French use intonation somewhat differently – a rising tone toward the end of the sentence typically indicates a question, while a steady or falling tone signals a statement.

Despite being classified as a Germanic language, English follows a distinct pattern in this regard. Although intonation plays an important role in conveying emotion and emphasis, the interrogative nature is conveyed by syntactic structure such as word placement (e.g., *are we clear?* versus *we are clear*) and the use of auxiliary verbs (*do you know?*). This reliance on structural differences can pose a challenge for non-native speakers often default to using native intonation, often resulting in confusing sentences like "*you are coming?*". Such differences can easily lead to misunderstandings – the confusion created by the sentence "we are clear for takeoff?" can have dire consequences. Awareness of these differences and adherence to standard phraseology are, therefore, essential for preserving safety.

- **Phonetic Differences in Critical Terminology**: Variations in vowels and consonant pronunciation can lead to confusion between similar-sounding words or numbers. For instance, the number "two" and the preposition "to" can be easily confused in phrases such as "Descend to two thousand feet." This can be misinterpreted as "Descend to twenty-two thousand feet," potentially leading to a serious altitude deviation. Similar phonetic overlaps occur with the words "four" and "for," or "nine" and the German word *nein* (meaning "no"), which can be particularly problematic in airspaces where operators routinely use plain language for communicating with flight controllers. These phonetic overlaps highlight the need for standardized phraseology and the use of phonetic alphabets and number pronunciation protocols (e.g., "niner" instead of "nine") to reduce ambiguity and ensure safety.
- **Rapid Speech and Insufficient Pausing**: The speed and rhythm of speech can significantly impact communication effectiveness. It is not uncommon for pilots and air traffic controllers to speak at a high rate, often with insufficient pausing between words, when operating in the congested airspace of major international airports. This rapid-fire delivery is frequently driven by competition for limited time on busy communication channels. It may also stem from a widespread – though often misplaced – assumption that crews and controllers operating in such high-intensity environments are highly experienced and therefore capable of processing fast-paced exchanges without difficulty. However, excessive speech speed and lack of clear pauses can lead to miscommunication, especially for non-native English speakers or in high-stress situations where clarity is critical (Prinzo, O. V et al., 2006).

- **Use of Idiomatic or Colloquial Language**: The use of idiomatic or colloquial expressions is common in everyday communication between individuals who share a cultural background. Like many other aspects of culture, these expressions reflect shared historical influences and constitute metaphors and linguistic variations that developed organically through social interaction. These idioms are often opaque to those unfamiliar with the culture since very little relationship exists between the meaning conveyed and the actual words uttered (think "*bite the bullet*"). The ambiguity and lack of universal comprehension of expressions like "*Roll the trucks*" and "*Hold your horses*" make them inappropriate for use in aviation (since they can lead to delayed response times or incorrect actions, especially in critical situations.
- **Strong Accents**: According to a study by Barshi and Farris (2013), approximately 22% of miscommunication reports in ATC involve accent-related factors. Strong regional accents can disrupt aeronautical communications by distorting the way numbers, letters, and even standard aviation phraseology are perceived and understood. Non-standard intonation patterns and variations in syllable or word pronunciation – such as those commonly observed among Indian, Scottish, or Irish speakers – can increase cognitive load for the listener, leading to delayed response times. This is particularly critical in high-density airspaces or during stressful, abnormal events. The need for repetition and clarification due to accented speech can also congest frequencies and reduce overall communication efficiency, diminishing overall communication efficiency and potentially impacting safety. To mitigate these challenges, ICAO recommends training aimed at neutralizing strong accents and raising awareness of the need to moderate local speech variations for non-native listeners. However, as previously discussed, this is not a trivial task – modifying deeply ingrained speech patterns is complex, as these variations develop early in life and are integral to an individual's cultural identity.

5.4.2 Communicating Across Cultures: Strategies and Best Practices

Having examined the threats and challenges in multicultural aviation communication in aviation, we now turn to practical solutions. Given the complexities of language, cultural differences, and their impact on communication, no single approach can fully resolve these issues. The following paragraphs discuss approaches that have been proven effective in mitigating the effects of cultural differences in aviation communication. These include training interventions designed to improve proficiency in aviation English (and English, in general), fostering cultural awareness, and increasing exposure to phonetic differences and idiomatic language. At the regulatory and organizational levels, enhancing standardized phraseologies, updating operator proficiency requirements, and implementing technical solutions as they become available – such as datalink systems for communication and surveillance (e.g., CPDLC and ADS B/C systems) – are essential steps toward greater efficiency and clarity in multicultural environments.

Promoting operator proficiency in standardized phraseology plays a crucial role in mitigating the effects posed by cultural diversity in international aviation. This extends beyond commercial aviation – standardized phraseology is fundamental to many globalized activities that rely upon clear, unambiguous communication. However, any training in standardized phraseology should also recognize and highlight its inherent limitations – there will always be situations where a standardized phraseology simply does not exist! Some interventions that can be effective in promoting the use of standardized phraseology, as well as some of their potential shortcomings, are:

- **Comprehensive and Operationally Relevant Training Programs**. While initial pilot or ATCO training places particular emphasis on foundational workshops, such as those focused on pronunciation, clarity, and accent reduction, recurrent training often fails to maintain this focus, with biannual simulator details offering little in the way of training communication techniques in multicultural ATC environments. This is at odds with the intention of the EBT/CBTA schemes that include communication as part of their non-technical skills. Regular, scenario-based training focusing on the use of correct phraseologies, along with the use of synthetic or prerecorded communications, could offer some mitigation to the limitations inherent in simulator training. Additionally, instructor training in creating realistic ATC environments (including the use of idiomatic language) could also be of some utility, although this might prove difficult in practice – the role-playing skills required for this might be beyond the typical simulator instructor. As we will see later, technological solutions, using artificial intelligence to generate ATC communications in simulated environments, are in development and may provide a solution to this problem. Finally, training scenarios that develop communication skills under high-stress situations might prove effective against teamwork and communication breakdowns frequently observed under these conditions (Kanki, 2019).
- **Crew Resource Management (CRM) Training**. CRM training programs have included modules on communication and culture since their inception in the late 1980s. Early courses often focused on psychological and management theories, which created friction amongst pilots who saw little relevance to their day-to-day work (Helmreich et al., 1999). Modern CRM training typically features presentations and group activities on basic communication theory, barriers to communication, and strategies for increasing efficiency. Most CRM modules on communication emphasize clarity, structure, and discipline as effective means of ensuring that critical information is conveyed accurately and efficiently, reducing ambiguities integral to multicultural interactions (Peksatici, 2018). Similarly, most CRM courses include modules on cultural differences, where basic theories of culture are introduced and best practices for intercultural communication are presented. Finally, it is very common for CRM syllabi to discuss case studies

of past incidents that highlight the risks associated with miscommunication. CRM courses can be particularly effective for presenting non-technical topics that are, as discussed, difficult to train and assess using traditional techniques designed for technical tasks. Nevertheless, it is important for course developers to balance between making the content comprehensive and informative while keeping it engaging for an audience that prioritizes practical application over academic discourse. Furthermore, although communication and culture modules are comprehensively discussed in isolation, it is seldom the case that a connection between the two is highlighted – there is rarely an integrated module on *intercultural communication*. Finally, an additional barrier to CRM effectiveness is its deep roots in Western cultural norms (Helmreich et al., 1999) – which may limit their adoption and transferability to line operations for crews from different cultural backgrounds.

- **Adherence to *Readback* and *Hearback* procedures**. Structured readback protocols, as mandated by ICAO, are essential in ensuring that instructions are both correctly understood and properly executed. These procedures require pilots to repeat back clearances and other safety-critical information to air traffic controllers, who then listen carefully (the "hearback") to confirm that the message was received accurately. This closed-loop communication system is important to ensure that the necessary *feedback* component of the communicative process (see Figure 5.1) is present despite the absence of confirmatory, non-verbal cues normally present in verbal communication. Readback and Hearback procedures are introduced to pilots and air traffic controllers early in their training and are reinforced throughout their careers. Their consistent application is expected in all but the most lenient air traffic control environments, serving as a fundamental safety net even when language barriers or heavy accents are present. Regular practice of these procedures not only reinforces communication discipline but also helps develop a shared expectation of clarity and accuracy between crews and controllers. In multicultural aviation settings, where crews and controllers may not share the same first language, strict adherence to readback and hearback protocols becomes even more critical, ensuring that essential information is transmitted and confirmed without ambiguity.
- **Technical Solutions**. Several automated systems are becoming available for pilot training in ATC communication. Typical examples include ASTi's SERA (Simulated Environment for Realistic ATC), developed for the training of ATC controllers and pilots, and the Simulated ATC Environment – SATCE. These leverage the flexibility of computerized speech recognition and synthesis, as well as the recent availability of technologies that enable the development of large language models (LLMs) trained on real-world pilot and controller interactions. The result is systems that provide dynamic, life-like interaction with artificial intelligence (AI) agents that exhibit culturally appropriate linguistic and communication styles. This facilitates the development of flight training programs that free the instructor from their current role-playing duties and enables them to concentrate on their core training and assessment tasks. At the same time, instructors maintain

of intervening, if necessary, to provide instructions or talk privately with trainees (a requirement in some training scenarios) or with other instructors (Advanced Simulation Technology Inc., n.d.). From an operational perspective, datalink systems are becoming increasingly common, and are mandated, in several airspace environments (for instance, in Europe). A typical example of such a system is the Controller Pilot Datalink System (CPDLC) that has been in use for several years now. The provision of air traffic control through short written messages provides a number of benefits, like lower frequency congestion, reduced probability of miscommunication due to linguistic differences, elimination of call sign confusion, and others. Some skepticism exists, nevertheless, about the loss of situation awareness these systems can create, mostly in relation to the targeted nature of the transmissions – other crews operating in the same airspace will not be able to form a mental model of potential conflicts with traffic operating in the same airspace since ATC instructions to other aircraft will not be received.

- **ICAO Level 4: Structured Prerequisite for Global Communication Safety**

  As commercial aviation continues to evolve and globalize, the need for clear, consistent, and culturally adept communication becomes increasingly evident. This section has examined some of the intricate relationships between language, culture, and technology in a domain where the margin for error is exceptionally narrow, and consequences can be severe. While standardized English phraseology and regulatory frameworks provide a crucial foundation for mutual understanding, they alone are insufficient to address the nuanced challenges posed by linguistic variability, divergent cultural norms, and human limitations under stress. Miscommunication, often rooted in phonetic discrepancies, accent interference, and culturally distinct communicative patterns, remains a persistent contributor to incidents and inefficiencies within the aviation system.

  Addressing the challenges of multicultural communication in aviation necessitates a comprehensive approach that integrates structured language training, cultural literacy, regulatory intervention, and technological advancements. Developing proficiency in standardized phraseology through programs that emphasize communication techniques in culturally diverse environments is fundamental during initial pilot and ATCO training. Then, a continued emphasis on practical, scenario-based communication training, supported by Crew Resource Management (CRM) modules and robust readback–hearback protocols, can further foster disciplined yet adaptive operational practices. The emergence of intelligent simulation platforms and datalink systems presents new opportunities to enhance both linguistic resilience and situational awareness in multicultural contexts. Ultimately, safeguarding aviation communication requires an ongoing commitment – not only to technical precision but to cultivating the intercultural sensitivity and adaptability that define effective global collaboration. By recognizing communication as both a procedural and human-centric endeavor, the aviation community is better positioned to uphold safety, efficiency, and inclusivity in increasingly diverse skies.

5.5 THE ORGANIZATIONAL CULTURE: SAFETY NORMS AND CREW RESOURCE MANAGEMENT STRATEGIES

Daan Vlaskamp

While much of the focus in this book so far has been on cultural differences in terms of national culture, this is not the only cultural factor that can influence safety in aviation. Where national culture is related to the shared values and beliefs of a specific group of people within a nation, professional culture concerns the shared values and beliefs held by people with the same profession. For example, many pilots share a sense of professionalism and a love of aviation. In this subchapter, the focus will be on organizational culture. Although organizational culture is influenced by national culture, it is a distinctly separate concept. The concept gained popularity in the 1980s. Barney (1986) defines organizational culture as "a complex set of values, beliefs, assumptions, and symbols that define the way in which a firm conducts its business." A subset of organizational culture is safety culture, which encompasses shared values, beliefs, and assumptions, specifically concerning safety. Several major disasters in the 1980s exposed organizational deficiencies related to safety and led to the rise of interest in the concept of safety culture in high-risk organizations. In the Challenger space shuttle disaster in 1986, for example, schedule pressure, poor communication, deficiencies in safety reporting, and the normalization of technical anomalies led to the dismissal of engineers' warnings against launching in cold weather, with catastrophic results. In the Chernobyl nuclear accident, also in 1986, factors such as fear of potential repercussions for decision-makers, structural secrecy, a focus on production, and a strong hierarchy contributed to the catastrophic reactor explosion and hampered the emergency response. Both disasters shifted the focus of accident investigations from blaming the human operator, to examining broader systemic factors, and helped put organizational factors, and safety culture in particular, firmly in the spotlight.

Thirty-two years later, in October 2018, Lion Air Flight 610 crashed into the Java Sea just minutes after takeoff from Jakarta, killing all 189 people on board. Less than five months later, Ethiopian Airlines Flight 302 suffered a near-identical fate, crashing shortly after takeoff from Addis Ababa and claiming 157 lives. Both aircraft were brand-new Boeing 737 MAX jets. Both incidents put the focus on safety culture again. This time, fingers pointed to the safety culture at the aircraft manufacturer: Boeing. At the heart of the two accidents was a new flight control system: the Maneuvering Characteristics Augmentation System (MCAS). It was introduced to improve the aircraft's stall characteristics. Information on the system was removed from the airplane's pilot manuals with FAA approval. One of the reasons for the removal of MCAS information from the manuals was economic pressure: conversion training from previous 737 models needed to be kept as short as possible. This way, existing customers of the 737 would be enticed to remain loyal instead of switching to the competition.

More incidents with Boeing aircraft followed, such as the blowout of a door plug on another 737 MAX. A Federal Aviation Administration (FAA, 2024) report harshly criticized Boeing and stated that cost-cutting measures, confusion among employees regarding safety procedures, and deficiencies in safety reporting all contributed to critical oversights. Many of the recommendations made by the FAA in the report raised concerns about Boeing's safety culture, which the FAA described as the "single most important influence on the management of safety."

Research into safety culture and the related concept of safety climate – a more surface-level snapshot of how safety is perceived and practiced in an organization –s demonstrates that a positive safety climate improves safety performance (Casey et al., 2017). In a strong safety culture, where safety is embedded into every level of an organization, workers are more motivated to work safely and to comply with safety regulations and procedures, leading to fewer incidents and accidents. They are better equipped with more knowledge on how to work safely and have the necessary tools to do so. Also, employees are more motivated to report safety issues. This has been shown to have a positive impact on the number of safety occurrences. However, fostering a strong safety culture has its challenges and takes time. It also requires a substantial financial investment, along with a genuine commitment from management. The benefits of investing in safety culture will not immediately be clear, as change takes time and actions needed to establish such a safety culture often clash with economic interests. Geert Hofstede, the famous cultural anthropologist, emphasized the importance of changing structures and systems, as changing culture by just focusing on employee behavior is almost impossible (Hofstede, 1991).

James Reason (1997) suggested that safety culture consists of five dimensions: an Informed Culture, a Reporting Culture, a Learning Culture, a Just Culture, and a Flexible Culture. These dimensions still form the basis of many airlines' safety policies. An Informed Culture ensures that an organization systematically collects, analyzes, and shares safety data across all layers of the company. Apart from automatic data monitoring, one of the ways in which data is gathered is through voluntary safety reporting by employees. In a positive Reporting Culture, employees feel confident that they can report safety concerns confidentially, without fear of reprisal. They should also be confident that the reports are used to enhance safety, and that reporting is thus useful and important. A strong Learning Culture ensures that the organization learns from its safety data and makes meaningful changes based on it. Creating a positive Learning Culture also involves motivating employees for continuous learning about safety. Another important aspect of safety culture is Just Culture. When reporting safety concerns, employees should feel confident that there will be no negative consequences for them personally, even when admitting to what could be perceived as mistakes or unsafe behavior. This makes employees less likely to conceal problems for fear of negative consequences, leaving important safety issues hidden under the surface. Just Culture is the opposite of a Punitive Culture, where individuals are blamed and face reprisals. That does not mean there is no personal accountability in a Just Culture: reckless, criminal, or negligent behavior still carries consequences. The disciplinary policy should be made transparent and known to the employees. Finally, a Flexible Culture helps the organization adapt to changing circumstances and demands and display resilience in the face of adversity. In the

Chernobyl and Challenger disasters, these safety culture dimensions showed deficiencies. After the Challenger accident, NASA committed to changing its culture and making safety the number one priority. Likewise, after the FAA report criticizing Boeing's safety culture and the 737 MAX accidents, Boeing put a lot of effort into strengthening its safety culture by encouraging employees to speak up, providing training in Just Culture principles, and establishing a digital safety learning platform. The result was an increase in safety reports by employees.

Numerous studies have identified management commitment as the most important factor influencing safety performance (such as Bosak et al., 2013). Investing resources in a safe work environment, involving employees in safety matters, performing job design according to safety principles, maintaining an ongoing safety dialogue, and, most importantly, putting words into action, are all examples of management practices that have a positive influence on safety. It is important for management to show a genuine commitment to safety. Recent research into organizational leadership in high-risk organizations has shifted from looking at the influence of leadership on productivity and efficiency to investigating the influence of leadership styles on safety outcomes. The concept of safety leadership (Flin & Agnew, 2018) emphasizes a style of leadership that motivates employees to prioritize safety in their work and has been shown to be beneficial to safety.

Pilots operate relatively autonomously when they are flying a plane. They do not interact with management daily, in the way that office workers who share a workspace with their immediate superiors do. This means that opportunities to strengthen safety culture and show management commitment to safety during training events and other such occasions become even more important. As well as mandatory simulator training and checking, pilots must undergo Crew Resource Management (CRM) training at regular intervals. The concept of CRM was developed in the late 1970s. In the early days of aviation, accidents often had technical causes. As technology advanced and aircraft became more and more safe, "human factors" contributing to aviation accidents became more apparent. Several high-profile crashes, such as the collision of two Boeing 747s on the runway of the airport of Tenerife, where communication difficulties and cockpit hierarchy played a role, put the spotlight on the human operator. CRM training initially focused on improving interpersonal skills (such as assertiveness), based on practices from management training in the corporate world and psychology. Now, CRM has gradually evolved to include the wider team outside the cockpit (e.g., cabin crew) and shifted to a more systematic approach. CRM training for airlines is mandated by law, and topics are prescribed by EASA and the FAA. One of the mandatory topics is the company's safety culture.

The way CRM is trained varies according to the airline. Most airlines use classroom training, while others lean more heavily on e-learning. CRM is also usually integrated into simulator training. Some of the training can be combined with other (non-flying) staff, which is known as joint CRM training. The CRM training environment is a place where company norms, attitudes, and beliefs can be communicated to and discussed with the wider workforce. The use of case studies and discussing incidents and accidents that have occurred in the company can be a powerful tool for establishing a Learning Culture. Through training and sharing safety knowledge, attitudes can be reinforced and crew members inspired and supported in working safely. By emphasizing the concept of Just Culture and the benefits of reporting, pilots can be motivated to contribute to a positive and effective Reporting Culture. Sharing success stories and discussing improvements made as a result of pilot reporting helps convey the message that management is committed to safety and that reporting is useful. Expectations of acceptable and unacceptable behavior related to safety can be made clear to employees to avoid uncertainty and improve compliance. Although the benefits of CRM are hard to quantify, several success stories in aviation have been ascribed to the successful application of CRM. The "Miracle on the Hudson" in 2009, where an aircraft successfully carried out an emergency water landing following a double engine failure, and Quantas Flight 32 in 2010, which landed successfully after suffering an uncontained engine failure after take-off that damaged critical systems and triggered dozens of alerts, are two examples of these.

Other useful training topics related to safety culture that can be discussed in training include hierarchy, operational drift, operational pressure, the importance of Standard Operating Procedures (SOPs), risk-taking, and team dynamics. Operational drift refers to the gradual and often unnoticed shift from baseline performance toward less safe practices. This usually begins with small deviations from established norms, which often go unchallenged as they remain consequence-free due to built-in safety margins. Over time, these deviations can become routine. This is known as "normalization of deviance." Such drift is frequently driven by operational pressures, whether externally imposed or self-inflicted, for example, the urge to stay on schedule, save fuel, or meet passenger expectations. If not identified and addressed, these seemingly minor deviations can erode safety margins and compromise overall safety. Economic pressures can have a strong negative impact on safety culture. Dekker (2017) states that systems like aviation do not exist to be safe; they exist to make money. This inherent "goal conflict" creates pressure on frontline staff. There is also the risk of mixed messages by management: paying lip service to safety, but simultaneously emphasizing the importance of making money, flying on time, or prioritizing passenger service.

Psychological safety is essential for a positive safety culture, according to the European Cockpit Association (ECA). Psychological safety, a concept first popularized by Amy Edmondson, refers to a climate where everyone feels comfortable expressing their opinions, questioning decisions, or reporting concerns without fear. It is an important concept, both inside the cockpit and on a company level. It highlights that creating the conditions where people feel safe to speak up does not only require effort from the individual who does so, but also commitment from the leadership in setting those conditions.

Deficient safety culture due to financial pressures and a lack of psychological safety can impact safety (Agenzia Nazionale per la Sicurezza del Volo, 2008). The captain of a Boeing 737 from a low-cost airline had, only days before the flight, lost his young son after a distressing period of illness. In fear of potential repercussions were he to take an extended leave of absence, he reported for duty despite his mental state following personal tragedy, and did not report the death to his Chief Pilot. While on final approach for Rome Ciampino Airport, the aircraft encountered severe turbulence and rain associated with thunderstorm activity. The flight diverted to Rome Fiumicino, but on the second approach, the aircraft ended up dangerously low, finally diverting again to Pescara, where it eventually landed safely. Both approaches in Ciampino and Fiumicino were unstable, with the captain showing signs of loss of situation awareness, cognitive overload, and impaired decision-making. These were likely exacerbated by the pilot's mental state and fatigue. The captain did not report the incident to the airline, again fearing reprisal.

Other forms of cultural differences can stem from individuals having absorbed different organizational norms and values over time. One example of these that has been shown to be particularly strong and enduring is the culture inculcated during training. This is one of the reasons why the role of instructors is seen to be so influential in establishing habits and safety culture. A captain and copilot can have a different training background (military versus civilian) or have spent years working in a different company (such as low-cost, legacy, or cargo) with a different organizational culture, even though they share the same national cultural background. Although the professional culture in aviation is relatively homogenous, these differences can influence everything from the way procedures are viewed to the authority gradient in the cockpit and can lead to misunderstandings and even conflicts, making them, if well established, even more influential over workplace behaviors than national culture. Emotional Intelligence is a skill that is more and more important in a globalized environment such as aviation. Developing Emotional Intelligence can mitigate the negative effects of cultural differences.

Training should not be used to offload organizational responsibility for safety issues on to frontline staff. As an example, although educating staff about fatigue and encouraging them to opt out of a flight when fatigued is important, management has an even greater responsibility to avoid scheduling that could lead to extreme fatigue. In a similar vein, emphasizing adherence to procedures is less effective in an environment where workers operate under strong operational or economic pressures. In the force-field between safety and efficiency, messages from management can send mixed signals, undermining the effectiveness of training. CRM training should never be a substitute for systemic safety measures. To be effective, training should also be tailored to the specific safety needs of the organization. Instead of relying on off-the-shelf training programs, a thorough training needs analysis should be carried out before developing training content.

Future developments in aviation will influence safety culture and CRM training. The rapid development of new systems such as AI is likely to fundamentally change the role of pilots in the cockpit and present new challenges in dealing with automation. The pilot's role will move further from active operator to system supervisor. The drive toward single-pilot operations by aircraft manufacturers also raises new challenges in terms of crew dynamics. It is likely to impact traditional CRM concepts as the "crew" becomes dispersed and expanded with ground-based actors, such as "remote pilots" assisting from a distance in case of emergency or during approach and landing. The multi-pilot model of a backup human resource in the cockpit will be absent, and the sole remaining pilot will have to rely mainly on himself. In this environment, the importance of continuous investment in a positive safety culture is even greater, driven by the need to align even more diverse parties with different interests and overcome additional communication and cultural barriers. As the 737 MAX incidents show, the success of technological advancements still depends on continuous investment in the human component. A strong safety culture will remain an essential part of that.

5.6 CONCLUSION

Dimitrios Ziakkas

Multinational flight crews are no longer an exception in international aviation – they are the standard. In this increasingly interconnected environment, the ability of flight crews to collaborate across cultural boundaries is not simply a desirable trait but a vital safety competence. As this chapter has demonstrated, the benefits of cultural diversity – enhanced problem-solving, broader situational awareness, and adaptive resilience – can only be fully realized when supported by inclusive training frameworks, culturally attuned CRM practices, and regulatory alignment.

However, diversity in the cockpit also introduces complexity, particularly in high-stakes operational settings where miscommunication or culturally influenced hesitation can have immediate consequences. The role of regulatory bodies and training organizations is therefore not just to standardize procedures, but to cultivate the competencies that allow for flexibility within that standardization. CBTA and EBT offer promising pathways, especially when paired with robust instructor development and ongoing human factors research that respects cultural nuance.

Looking ahead, emerging technologies such as AI-driven decision aids, biometric monitoring, and intelligent automation will continue to reshape cockpit ergonomics and communication flows. Yet, these tools must be deployed with an understanding of how human culture mediates technological interaction. Ergonomics is not only about physical layout – it is about psychological accessibility, cognitive load, and interpersonal trust.

Ultimately, the safety and efficiency of multinational crews depend on a collective commitment to learning, adaptation, and empathy. *The flight deck must be recognized as a dynamic social space, not merely a technical one.* By embracing cultural intelligence as an integral part of operational competence, the aviation industry can transform diversity from a challenge into an enduring source of strength.

REFERENCES

Advanced Simulation Technology Inc. (n.d.). Intelligent ATC agents augmenting Simulator-Based, Military flight training. https://www.asti-usa.com/progexs/military/serappt.html#:~:text=Automated%20Entities%20That%20Role%2DPlay,or%20text%20with%20other%20instructors

Agenzia Nazionale per la Sicurezza del Volo (ANSV). (2008). Final report: Serious incident to B737-800, registration EI-DAV, on approach to Rome Ciampino and Fiumicino airports, September 7th, 2005. ANSV.

Al-Wardi, Y. (2017). Arabian, Asian, western: A cross-cultural comparison of aircraft accidents from human factor perspectives [Article]. *International Journal of Occupational Safety and Ergonomics*, 23(3), 366–373. https://doi.org/10.1080/10803548.2016.1190233

Atchley, P., Shi, J., & Yamamoto, T. (2014). Cultural foundations of safety culture: A comparison of traffic safety culture in China, Japan and the United States [Article]. *Transportation Research Part F: Traffic Psychology and Behaviour*, 26(PB), 317–325. https://doi.org/10.1016/j.trf.2014.01.004

Barney, J. B. (1986). Organizational culture: can it be a source of sustained competitive advantage?. *Academy of Management Review*, 11(3), 656–665.

Barshi, I., & Farris, C. (2013). *Misunderstandings in ATC communication: Language, cognition, and experimental methodology*. Ashgate

Baumann, G. (1996). *Contesting culture : discourses of identity in multi-ethnic London*. Cambridge University Press.

Ben-Saed, M., & Pilbeam, C. (2022). The effect of an embargo, sanctions and culture on safety climate: A qualitative view from aviation maintenance in the MENA region [Article]. *Journal of Safety Research*, 81, 259–269. https://doi.org/10.1016/j.jsr.2022.03.003

Bosak, J., Coetsee, W. J., & Cullinane, S. J. (2013). Safety climate dimensions as predictors for risk behavior. *Accident Analysis & Prevention*, 55, 256–264.

Casey, T., Griffin, M. A., Flatau Harrison, H., & Neal, A. (2017). Safety climate and culture: Integrating psychological and systems perspectives. *Journal of Occupational Health Psychology*, 22(3), 341.

Cathay Pacific (2025) Our commitment to diversity and inclusion. cathaypacific.com. https://www.cathaypacific.com/cx/en_HK/about-us/diversity-and-inclusion.html

Chan, W. T. K., & Li, W. C. (2021). Culture's consequences on the categorisation of causal factors in aviation accident reports. In *Lecture Notes in Computer Science* (including subseries Lecture Notes in Artificial Intelligence and Lecture Notes in Bioinformatics) (Vol. 12767 LNAI, pp. 19–27). https://doi.org/10.1007/978-3-030-77932-0_2

Cox, T. H., & Blake, S. (1991). Managing cultural diversity: Implications for organizational competitiveness. *Academy of Management Perspectives*, 5(3), 45–56.

Dekker, S. (2017). *The field guide to understanding 'human error'*. CRC Press

EASA. (2020). *Acceptable means of compliance and guidance material to Part-FCL*. European Union Aviation Safety Agency. https://www.easa.europa.eu/en/document-library/acceptable-means-of-compliance-and-guidance-materials/part-fcl

EASA. (2021). Guidelines for Competency-Based Training and Assessment (CBTA). https://www.easa.europa.eu/en/newsroom-and-events/news/competency-based-training-and-assessment

Enomoto, C. E., & Geisler, K. R. (2017). Culture and plane crashes: A cross-country test of the Gladwell hypothesis [Article]. *Economics and Sociology*, 10(3), 281–293. https://doi.org/10.14254/2071-789X.2017/10-3/20

Federal Aviation Administration. (2024). Section 103 Organization Designation Authorizations (ODA) for Transport Airplanes Expert Panel Review Report. U.S. Department of Transportation. https://www.faa.gov/newsroom/Sec103_ExpertPanelReview_Report_Final.pdf

Flin, R., & Agnew, C. (2018). *Human factors in safety management: Safety culture, safety leadership and non-technical skills*. CRC Press

Gudykunst, W. B., & Mody, B. (Eds.). (2002). *Handbook of international and intercultural communication*. Sage.

Hazrati, A. (2015). Intercultural communication and discourse analysis: The case of Aviation English. *Proceedings of 2nd Global Conference on Linguistics and Foreign Language Teaching*. https://doi.org/10.1016/j.sbspro.2015.06.035

Helmreich, R., Anca, J., & Kanki, B. G. (2010). The future of CRM. In *Crew resource management* (pp. 493–500). Academic Press

Helmreich, R. L., Merritt, A. C., & Wilhelm, J. A. (1999). The evolution of Crew Resource Management training in commercial aviation. *International Journal of Aviation Psychology*, 9(1), 19–32.

Hodgson, A., Siemieniuch, C. E., & Hubbard, E. M. (2013). Culture and the safety of complex automated sociotechnical systems. *IEEE Transactions on Human- Machine Systems*, 43(6), 608–619. https://doi.org/10.1109/THMS.2013.2285048. Article 6650057.

Hoecklin, L.A. (1995). *Managing cultural differences: Strategies for competitive advantage/ Lisa Hoecklin*. Addison-Wesley

Hofstede, G. (1991). *Cultures and organisations: Software of the mind.* McGraw-Hill

Hofstede, G. (2003). *Culture's consequences: Comparing values, behaviors, institutions, and organizations across nations* (2nd ed.). Sage Publications

Hofstede, G., Hofstede, G. J., & Minkov, M. (2010). *Cultures and organizations: Software of the mind* (3rd ed.). McGraw-Hill.

Holliday, A. (2001). Culture as constraint or resource. in: Gray, K., Ansell, M., Cardew, S. and Leedham, M. (ed.) *The Japanese Learner: Context*, Culture and Classroom Practice Oxford Oxford University Press.

ICAO. (2013). Manual of Evidence Based Training. Doc 9995. AN/497

ICAO. (2021). The annual ICAO safety report. International Civil Aviation Organization. https://www.icao.int/safety/Documents/ICAO%20Safety%20Report%202021%20Edition.pdf

Jing, H. S., & Batteau, A. (2015). *The dragon in the cockpit: How Western aviation concepts conflict with Chinese value systems* (1st ed.). Routledge.

Jing, H. S., Lu, C. J., & Peng, S. J. (2001). Culture, authoritarianism and commercial aircraft accidents. *Human Factors and Aerospace Safety - An International Journal*, 1 (4), 341–359.

Kanki, B. G. (2019). Communication and crew resource management. In *Crew resource management* (pp. 103–137). Academic Press.

Klein, H. A. (2005). Cultural differences in cognition: Barriers in multinational collaborations. In H. Montgomery, R. Lipshitz, & B. Brehmer (Eds.), *How professionals make decisions* (pp. 243–253). Lawrence Erlbaum.

Kwon, Y. (2019). A review on the national culture and its effects on the transportation safety perspectives. *TransNav: International Journal on Marine Navigation and Safety of Sea Transportation*, 13(2), 275–281. https://doi.org/10.12716/1001.13.02.02

Leveson, N. (2004). A New Accident Model for Engineering Safer Systems. *Safety Science*, 42(4), 237–270

Li, W. C., Harris, D., Li, L. W., & Wang, T. (2009). The differences of aviation human factors between individualism and collectivism culture. In *Human-Computer Interaction. Interacting in Various Application Domains: 13th International Conference, HCI International 2009*, San Diego, CA, USA, July 19–24, 2009, Proceedings, Part IV 13 (pp. 723–730). Springer Berlin Heidelberg.

Liao, M. Y. (2015). Safety culture in commercial aviation: differences in perspective between Chinese and western pilots [Article]. *Safety Science*, 79, 193–205. https://doi.org/10.1016/j.ssci.2015.05.011

Merrit, A. C., & Helmreich, R. L. (1996). Human factors on the flight deck: The influence of national culture. *Journal of Cross-Cultural Psychology*, 27(1), 5–24.

Meyer, E. (2016). *The culture map (INTL ED): Decoding how people think, lead, and get things done across cultures*. PublicAffairs.

Noort, M. C., Reader, T. W., & Gillespie, A. (2021). Safety voice and safety listening during aviation accidents: Cockpit voice recordings reveal that speaking-up to power is not enough. *Safety Science*, 139, Article 105260. https://doi.org/10.1016/j.ssci.2021.105260

Page, S.E. (2017). *The diversity bonus: how great teams pay off in the knowledge economy*. Princeton University Press.

Peksatici, O. (2018). Crew resource management (CRM) and cultural differences among cockpit crew - the case of Turkey. *Journal of Aviation/Aerospace Education & Research*, 27(2), 1–39. https://doi.org/10.15394/jaaer.2018.1742

Peterson, M. F. (2004). Review of Culture, Leadership and Organizations: The GLOBE Study of 62 Societies, by R. J. House, M. Javidan, P. W. Dorfman, V. Gupta, & Paul J. Hanges. *Administrative Science Quarterly*, 49(4), 641–647. https://www.jstor.org/stable/4131493

Pratama, G. B., & Caponecchia, C. (2025). Examining the influence of national culture on aviation safety: A systematic review. *Journal of Safety Research*, 92, 317–330.

Pratama, G. B., Widyanti, A., & Sutalaksana, I. Z. (2018). Incorporating Hofstede' National Culture in Human Factor Analysis and Classification System (HFACS): Cases of Indonesian Aviation Safety. *MATEC Web of Conferences*. https://doi.org/10.1051/matecconf/201815401063

Reader, T. W., Noort, M. C., Shorrock, S., & Kirwan, B. (2015). Safety sans fronti`eres: An international safety culture model [Article]. *Risk Analysis*, 35(5), 770–789. https://doi.org/10.1111/risa.12327

Reason, J. (1997) *Managing the Risks of Organizational Accidents*. Routledge.

Ritsher, J. B. (2005). Cultural factors and the international space station. *Aviation, Space, and Environmental Medicine*, 76(6, Sect2, Suppl), B135–B144.

Schiffrin, D., Tannen, D. & Hamilton, H.E. (eds.) (2001). The handbook of discourse analysis. *Journal of Linguistics*, 40(1), 192–198.

Seva, R. R., Gutierrez, A. M. J. A., Duh, H.-B.-L., & Chong, J. (2007). An evaluation of CRM attitudes for Filipino pilots in four Phillipine aviation companies. *The International Journal of Aviation Psychology*, 17(3), 285–298. https://doi.org/10.1080/10508410701343532

Soeters, J. L., & Boer, P. C. (2000). Culture and flight safety in military aviation [Article]. *International Journal of Aviation Psychology*, 10(2), 111–133. https://doi.org/10.1207/S15327108IJAP1002_1

Stancu, G. (2013). *The impact of culture in flight safety*. Romanian National Defense University, Regional Department of Defense Resources Management Studies.

Straub, D., Loch, K., Evaristo, R., Karahanna, E., & Srite, M. (2002). Toward a theory-based measurement of culture. *Journal of Global Information Management (JGIM)*, 10(1), 13–23.

Strauch, B. (2010). Can cultural differences lead to accidents? Team cultural differences and sociotechnical system operations. *Human Factors*, 52(2), 246–263.

Tear, M. J., Reader, T. W., Shorrock, S., & Kirwan, B. (2020). Safety culture and power: Interactions between perceptions of safety culture, organisational hierarchy, and national culture [Article]. *Safety Science*, 121, 550–561. https://doi.org/10.1016/j.ssci.2018.10.014

Trompenaars, F., & Hampden-Turner, C. (2020). *Riding the waves of culture: Understanding diversity in global business* (4th ed.). McGraw-Hill

Urban, J. M., Weaver, J. L., Bowers, C. A., & Rhodenizer, L. (1996). Effects of workload and structure on team processes and performance: Implications for complex team decision making. *Human Factors*, 38, 300–310.

Zhu, S. X., & Ma, W. L. (2015). Culture's influence on cockpit communication. *Proceedings of the International Conference on Management, Computer and Education Informatization*. https://doi.org/10.2991/mcei-15.2015.109

6 Implementing Cultural Intelligence in Training

6.1 INTRODUCTION

Dimitrios Ziakkas

In the evolving landscape of global aviation, the integration of Cultural Intelligence (CQ) into Crew Resource Management (CRM) training has emerged as a critical priority. While CRM has traditionally focused on teamwork, communication, and decision-making, the increased diversity of crews – fueled by globalization, international partnerships, and cross-border labor mobility – demands a training paradigm that explicitly incorporates intercultural dynamics. In practical terms, embedding CQ within CRM requires rethinking the way modules are structured. Communication modules may include exercises that contrast direct and indirect communication styles or encourage explicit phraseology in high-context cultures. Decision-making scenarios should incorporate culturally situated dilemmas, such as how hierarchy and power distance might affect assertiveness in safety-critical discussions. To ensure relevance and operational rigor, learning objectives must clearly articulate the link between CQ awareness and measurable safety outcomes. Interactive learning experiences (e.g., case-based learning, structured reflection, and intercultural role-play) are key to developing these competencies. Through post-exercise reflection, learners are encouraged to consider how motivational CQ (the willingness to engage with culturally different others) influences conflict resolution efficacy.

This chapter explores the theoretical foundation, instructional design, practical application, and assessment mechanisms necessary to embed CQ within CRM training effectively. In doing so, it draws upon global regulatory frameworks, academic research, and operational best practices to provide a roadmap for competency-based, culturally intelligent aviation training.

6.2 DEVELOPING TRAINING MODULES

Dimitrios Ziakkas, Anastasios Plioutsias, Dimitra Synodinou,
Aspasia Argyrou and Ioanna Lekea

The International Civil Aviation Organization (ICAO) and International Air Transport Association (IATA) have increasingly recognized the human-centric dimensions of safety performance. ICAO's Global Aviation Safety Plan and Human Factors guidelines underscore the importance of harmonizing human performance factors within multinational crews (ICAO, 2013). Evidence-Based Training (EBT) and Competency-Based Training and Assessment (CBTA) frameworks, supported by both ICAO and EASA, offer a pathway toward integrating behavior and culture.

DOI: 10.1201/9781003672029-8

This section follows the ICAO-endorsed **ADDIE model (i.e., Analysis, Design, Development, Implementation, and Evaluation**) to structure the integration of CQ into CRM training. This model ensures that training is systematic, learner-centered, and operationally relevant. Concurrently, we introduced **Kirkpatrick's four levels of training evaluation (i.e., Reaction, Learning, Behavior, and Results**) to provide a continuous feedback loop and empirical basis for improvement. The resulting framework supports evidence-based, culturally intelligent, and competency-driven crew performance aligned with ICAO (Doc 9995), FAA (AC 120-51E), EASA, and IATA standards.

- **Analysis**
  The analysis phase involves identifying operational needs, regulatory mandates, and cultural challenges that affect CRM outcomes. Airlines operating across multicultural environments report increased instances of miscommunication, hierarchical misalignment, and conflicting decision styles – factors that are known contributors to aviation incidents (Helmreich & Merritt, 2000). A thorough training needs analysis (TNA) must identify these gaps by reviewing incident reports, survey data, and post-flight debriefs.
- **Design**
  In this phase, learning objectives are aligned with both CRM competencies and CQ dimensions. For instance, CRM's "Communication" competency maps onto CQ's cognitive and behavioral domains. A training objective may read: "Demonstrate appropriate adaptation of communication style in culturally diverse cockpit teams." Moreover, **Kirkpatrick Level 1 (Reaction)** is addressed at this stage by designing content that is engaging, contextually relevant, and sensitive to participants' cultural backgrounds.
- **Development**
  Content development includes the creation of lesson plans, simulation scripts, and case-based scenarios. Interdisciplinary teams (e.g., comprising pilots, cabin crew, instructional designers, and organizational psychologists) should collaborate to ensure accuracy and relevance. **Kirkpatrick Level 2 (Learning)** is evaluated through formative assessments embedded within the module. For example, crews may complete pre- and post-training CQ self-assessments or scenario-based quizzes.
- **Implementation**
  Training delivery may involve classroom instruction, web-based modules, and full-motion simulator sessions. At this point, CRM instructors must be equipped to handle culturally nuanced dialogue and respond constructively to culturally derived misunderstandings. The followed **Kirkpatrick Level 3 (Behavior)** focuses on post-training transfer to the line environment. In summary, are crews demonstrating adaptive communication, inclusive decision-making, and mutual respect across cultural lines?
- **Evaluation**
  The final ADDIE phase includes summative evaluations such as instructor feedback, participant reflections, and operational performance metrics. Safety reports, Line Operations Safety Audits (LOSAs), and debriefing data

should be reviewed longitudinally. At **Kirkpatrick Level 4 (Results)**, metrics such as reduction in cultural miscommunication, improved CRM scores, and safety incident trends serve as indicators of training effectiveness.

6.2.1 Curriculum Design

As global aviation systems evolve into increasingly interconnected, multicultural environments, traditional CRM frameworks must adapt. CRM is not a mere vehicle for promoting communication and teamwork; today, it also serves as a critical tool for navigating the subtleties of intercultural interaction. Integrating CQ into CRM curricula enables aviation crews to move beyond mere technical proficiency, fostering deeper awareness of how cultural values, beliefs, and behaviors shape decision-making and teamwork in the cockpit.

The foundation of an effective CQ-integrated CRM curriculum begins with aligning the four core CQ dimensions (i.e., metacognitive, cognitive, motivational, and behavioral) with established CRM objectives such as communication, situational awareness, decision-making, and leadership (Ang & Van Dyne, 2008). For example, metacognitive CQ enhances situational awareness by encouraging reflective thought on how cultural assumptions influence interpretation, while behavioral CQ informs communication style adaptation in diverse crews. Literature review findings support the efficacy of CQ-informed training across multiple domains. Studies from high-risk industries such as healthcare and maritime operations demonstrate that teams with higher CQ show improved cooperation and reduced error rates in multicultural environments. In aviation, where safety hinges on teamwork and efficient communication, these findings underscore the operational necessity of culturally intelligent training. The TNA focuses on the following:

6.2.1.1 Alignment of CQ with CRM Learning Objectives

A successful integration of CQ into CRM must begin with a careful mapping of how each CQ dimension supports CRM objectives:

- **Metacognitive CQ** promotes critical reflection on how culture influences perception. In CRM, this enhances situational awareness and decision-making.
- **Cognitive CQ**, defined as knowledge about different cultures' norms and practices, informs crew members' understanding of diverse team expectations, supporting effective teamwork.
- **Motivational CQ**, or the interest and confidence in functioning in culturally diverse contexts, is essential in maintaining crew cohesion and reducing cultural avoidance during high-stress interactions.
- **Behavioral CQ**, the ability to adapt verbal and non-verbal actions, aligns with CRM's focus on communication clarity and team dynamics.

This alignment allows for deliberate pedagogical design, ensuring that each CRM competency is enriched by targeted CQ development.

6.2.1.2 Interdisciplinary Curriculum Integration

Integrating CQ into CRM is not merely an additive process – it requires interdisciplinary curriculum design that draws from fields such as psychological anthropology, intercultural communication, and applied ergonomics. Ang and Van Dyne's (2008) four-factor model provides a conceptual anchor, while practical training content is informed by ICAO's Human Factors Training Manual (Doc 9683) and EASA's guidelines on EBT (EASA, 2021).

A typical module sequence may include:

- **Theoretical input**, such as mini-lectures or readings introducing CQ concepts.
- **Case study analysis**, examining real-world flight deck incidents involving intercultural dynamics.
- **Simulated interactions**, such as mock pre-flight briefings with culturally diverse crews.
- **Assessment components**, including self-reflection exercises and peer feedback.

Curriculum designers must also be mindful of how learning is scaffolded. For instance, early modules might focus on self-awareness and cultural perception, while later modules demand more complex applications, such as culturally adaptive leadership during simulated emergencies.

6.2.2 Operational Relevance and Industry Standards

Adopting CQ into CRM training is increasingly a regulatory and operational imperative. ICAO (2013), in its recent discussions on global training harmonization, highlights the importance of embedding cultural competence within CRM programs to address the realities of multinational crews. The FAA's Advisory Circular 120-51E also encourages CRM instructors to be culturally sensitive, especially in training programs for international operators. EASA (2021) has begun integrating diversity awareness into instructor evaluation rubrics for EBT implementation. By aligning CQ-enhanced CRM programs with such regulatory guidance, organizations not only ensure compliance but also position themselves to improve operational performance.

Global aviation is not monolithic; operational environments vary widely depending on airline culture, geographical region, and mission type. Consequently, CRM programs must be adaptable to local contexts while maintaining global safety standards. This is particularly true in mixed environments such as ATC–pilot interactions or multicultural cabin crew teams. For instance, an operator flying long-haul routes between East Asia and Northern Europe may face communication challenges rooted in divergent cultural norms around authority and directness. A tailored module might incorporate case studies relevant to these regions, role-plays involving native phraseology differences, and scenario-based planning that emphasizes pre-emptive briefing practices. Some programs may go further by integrating **Diversity Intelligence (DQ)** components, extending beyond culture to include gender, age, neurodiversity, and professional backgrounds (Livermore, 2024).

6.2.3 Practical Exercises and Simulations

Simulation-based training is central to aviation competency development and offers a compelling means of cultivating CQ within CRM instruction. By immersing participants in controlled, yet emotionally and cognitively demanding intercultural interactions, role-play enables crews to experience first-hand the nuances and pressures of multicultural teamwork. When designed with intention and grounded in operational authenticity, these scenarios move beyond rote procedural training and facilitate the experiential learning essential for long-term behavioral change. A culturally intelligent CRM scenario might involve a crew comprised of individuals from high and low power-distance cultures navigating a time-critical in-flight system failure. The exercise could center on the tension between hierarchical deference and assertive communication. A first officer from a culture that emphasizes respect for seniority may hesitate to challenge the captain's course of action, despite recognizing a deviation from standard operating procedures. The resulting delay in addressing the issue presents a realistic opportunity to examine how motivational and behavioral CQ influence communication and decision-making under pressure. These scenarios also allow participants to explore the impact of differing cultural norms around non-verbal communication, emotional expression, and conflict resolution. For example, some cultures view emotional restraint as a mark of professionalism, while others expect visible emotional engagement as a sign of honesty and commitment. Without cultural awareness, these differences can be misinterpreted, leading to unnecessary friction or breakdowns in teamwork. Simulation provides a psychologically safe space to uncover these mismatches, reflect upon them, and rehearse adaptive strategies (Salas et al., 2020).

6.2.3.1 Scenario Development

Developing effective role-playing scenarios for CQ-CRM training requires careful attention to both instructional design and operational fidelity. A robust scenario development process typically includes:

- **Defined learning objectives** tied explicitly to CQ dimensions and CRM competencies.
- **Contextual realism**, ensuring that scenarios reflect actual operational challenges, such as long-haul crew changeovers or mixed-nationality ATC environments.
- **Cultural briefing materials**, outlining the national or organizational cultures involved and the potential value conflicts.
- **Facilitator guides** to structure observation and post-simulation debriefing.
- Instructors should also ensure that cultural representations within scenarios are sensitive and evidence-based. Overgeneralizations or cultural caricatures undermine the credibility of the training and risk reinforcing stereotypes rather than building competence.

6.2.3.2 Simulated Operational Contexts

CQ dimensions manifest differently depending on the complexity and emotional tenor of the operational context. Low-stakes simulations, such as pre-flight briefing exercises, are ideal for developing metacognitive and cognitive CQ, allowing participants to reflect on cultural expectations and plan interactions accordingly. High-stress scenarios – such as medical emergencies, unplanned diversions, or unstable approaches – are more effective for fostering behavioral and motivational CQ, as they require immediate and sustained intercultural collaboration under pressure.

For example, a simulator session involving a crew from multiple nationalities responding to an in-flight decompression can reveal subtle communication breakdowns caused by cultural variations in assertiveness or confirmation protocols. One crew member may interpret a direct command as overly aggressive, while another may misread a polite suggestion as non-urgent. These friction points provide valuable teaching moments when guided by skilled facilitators.

6.2.3.3 Interactive Role-Playing Methods

The efficacy of role-playing depends not just on scenario content, but on the methods used to facilitate engagement and reflection. Interactive methods that have shown effectiveness in CRM include:

- **Rotating roles**, where participants switch cultural or authority positions to experience different perspectives.
- **Stress inoculation scenarios**, which simulate high-pressure conditions requiring cultural agility.
- **Cross-cultural briefing and feedback loops**, ensuring that each exercise includes structured dialogue on how cultural expectations shaped communication and decisions.

These methods are reinforced by adult learning principles, which suggest that self-directed and experiential learning fosters greater retention and transfer to operational settings (Knowles et al., 2015).

6.2.3.4 Evaluation of Scenario Effectiveness

To ensure that simulation is not simply performative, robust evaluation strategies must accompany scenario execution. This includes:

- **Behavioral observation checklists**, aligned with CQ dimensions and CRM markers, allowing instructors to assess adaptability, empathy, and intercultural communication.
- **Structured participant feedback forms**, gathering data on perceived realism, emotional engagement, and learning outcomes.
- **Post-simulation self-reflection**, encouraging learners to articulate how they responded to cultural challenges and what they would do differently in future operations.

Data from these assessments can also be aggregated over time to track organizational trends in CQ development, supporting quality assurance and continuous improvement processes.

6.3 DEBRIEFING TECHNIQUES

Dimitrios Ziakkas, Konstantinos Pechlivanis and Ian Roth

The debriefing phase is the pedagogical heart of CQ-CRM training. It is in these reflective sessions – occurring after simulations or operationally inspired role-plays – that trainees begin to internalize lessons, identify cognitive and emotional responses, and develop practical strategies for improvement. Effective debriefing transforms culturally complex events into structured learning opportunities by foregrounding the cultural dimensions of decision-making, communication, and teamwork (Salas et al., 2020; Flin et al., 2013).

Culturally intelligent debriefing techniques differ from standard After Action Reviews in several key respects. They require facilitators to prompt participants not only about procedural accuracy or technical skill but also about **cultural interpretation and adaptation**. Questions such as "How did your cultural assumptions influence your response?" or "Did you notice any cultural differences in how your crewmates approached the task?" become central. These questions invite a deeper level of self-reflection and social learning than traditional performance feedback.

Moreover, effective debriefing focuses on **psychological safety** – the shared belief that interpersonal risk-taking will not result in embarrassment or retaliation. This is especially important when discussing culturally sensitive topics such as hierarchy, assertiveness, or communication norms. If not handled thoughtfully, these conversations can trigger defensiveness or reinforce existing cultural divides. Therefore, facilitators must structure discussions around curiosity, growth, and shared responsibility, rather than blame or critique.

6.3.1 CQ-Focused Debrief Templates

To operationalize CQ reflection, facilitators may use structured debrief templates incorporating indicators tied to each CQ dimension. Examples include:

- **Metacognitive CQ**: Did the crew anticipate cultural influences? Did they reflect during the event?
- **Cognitive CQ**: Was there an understanding of relevant cultural norms or expectations?
- **Motivational CQ**: Did participants demonstrate openness and persistence in intercultural interactions?
- **Behavioral CQ**: Were verbal and non-verbal communication styles adapted appropriately?

Templates can include Likert-scale ratings, open-ended questions, and team-based consensus forms. These instruments not only guide the facilitator but also serve **as learning tools** that reinforce key competencies through repetition and visibility.

Debriefing should also incorporate **peer feedback**, encouraging crew members to share how they experienced the interaction through their cultural lens. This pluralistic reflection helps reduce egocentric bias and enables collective learning from culturally diverse perspectives (Helmreich & Merritt, 2000).

6.3.2 Facilitator Competencies

The success of culturally intelligent CRM debriefing is highly dependent on the **skill, awareness, and credibility** of the facilitator. Instructors must possess not only a strong grasp of human factors and CRM but also:

- **Cultural humility** – an openness to learning from others' cultural perspectives without assuming expertise.
- **Cognitive empathy** – the ability to understand how culture influences others' perceptions and decisions.
- **Feedback skills** – the capacity to provide constructive, non-judgmental feedback that supports behavioral growth.
- **Conflict navigation** – the ability to defuse tension, maintain group cohesion, and redirect emotionally charged discussions toward productive insight.

Many of these competencies are not innate but can be developed through **train-the-trainer** programs that emphasize intercultural competence, facilitation theory, and reflective supervision. Airlines such as KLM and Qantas have incorporated such programs into their CRM instructor certification processes, acknowledging the expanded role of trainers in shaping inclusive and safe learning environments (IATA, 2020).

6.3.3 Learning Reinforcement

To ensure long-term effectiveness, CQ competencies must be reinforced beyond the immediate training context. Research in adult learning suggests that behavioral change is most sustainable when supported by repeated practice, reflection, and feedback loops (Knowles et al., 2015). This principle holds especially true in CRM, where high-pressure decision-making must be undergirded by deep-seated interpersonal fluency.

Some methods for reinforcing CQ learning include:

- **Follow-up debriefs** after real-world line flights, especially involving culturally diverse crews.
- **Microlearning modules** delivered digitally between simulator sessions, focusing on single CQ concepts (e.g., cultural display rules in communication).
- **Reflection journals** or peer discussion boards, allowing participants to track their intercultural development over time.
- **Integrated SMS feedback**, where observations of cultural miscommunication in operational reports are used (anonymously) as case studies in future training.

Embedding CQ into the organizational learning culture in this way ensures that CRM is not reduced to a classroom event but becomes part of the airline's operational DNA.

6.4 CASE VIGNETTES AND EXAMPLES

Dimitrios Ziakkas

The integration of Cultural Intelligence (CQ) into CRM training is no longer a theoretical exercise; it is a response to the lived realities of contemporary flight operations. Aviation history is rich with incidents and near-misses where culturally driven miscommunication, or mismatched expectations, played a decisive role – both positively and negatively. This section presents case vignettes that illustrate the tangible impact of CQ in action.

A well-known success case involved a long-haul flight operated by a Middle East carrier with a multinational crew composed of a Pakistani captain, a Romanian first officer, and Indian and Filipino cabin crew. During a mid-flight electrical malfunction, the first officer's recommendation to initiate an alternate checklist was initially met with silence. Rather than escalating tension, the first officer paused and calmly paraphrased the concern using ICAO-standard phraseology, adapting his tone and cadence to align with the captain's culturally preferred communication style. The captain, recognizing the seriousness of the situation, finally took the appropriate decision and course of action. Post-flight debriefings indicated that both behavioral and cognitive CQ were instrumental in ensuring effective communication, contributing to a successful resolution. Such examples reinforce the critical operational value of CQ. The ability to adjust communication style, to recognize when silence signals hesitation rather than agreement, and to bridge hierarchical expectations are not abstract competencies – they are flight-critical behaviors that can prevent errors and promote team resilience under pressure (Salas et al., 2020).

Another example comes from a European charter airline operating summer routes in the Mediterranean. The CRM training team introduced pre-flight cultural briefing cards as part of crew sign-on, summarizing key national tendencies related to communication, power distance, and conflict resolution. After three months, crew feedback and incident reports demonstrated a marked improvement in team coordination and reduced reports of interpersonal friction. This intervention, though simple, illustrates how proactive CQ application can yield measurable improvements in both operational performance and team morale.

6.4.1 Aviation Case Studies

To further ground the relevance of CQ in CRM, it is important to examine historical case studies where a lack of cultural awareness had operational consequences. One such case is the crash of Avianca Flight 52 in 1990, where delayed and indirect communication between a Colombian crew and American ATC contributed to fuel exhaustion and fatalities. The first officer's culturally influenced deference to authority and reluctance to use assertive phraseology ("we're running out of fuel") played a role in the failure to convey the criticality of the situation (Helmreich & Merritt, 2000).

Similarly, the Korean Air Flight 801 accident in 1997 highlighted the dangers of high power distance in cockpit communication. Investigations revealed that junior officers hesitated to challenge the captain's decisions, even when aware of serious navigation errors. These incidents led to regulatory and training overhauls emphasizing assertiveness, standard phraseology, and flattened hierarchies in flight deck interactions – many of which are now understood through the lens of CQ.

Such cases underscore the need for CRM training programs to explicitly address cultural dynamics – not as peripheral considerations, but as central elements of flight safety. They also validate the predictive value of CQ theory: that culturally mismatched communication styles, if not identified and adapted to, can become latent system threats.

6.4.2 Operational Outcomes

The benefits of CQ-integrated CRM training can be tracked through both qualitative feedback and quantitative performance indicators. Airlines that have implemented CQ-focused modules often report:

- Reduction in miscommunication-related safety reports.
- Improved team cohesion in line operations.
- Higher trainee satisfaction in simulator training.
- Increased confidence among crew in addressing cultural challenges.

Singapore Airlines and Emirates, for example, have incorporated structured CQ assessments into their recurrent training programs and observed more adaptive crew behavior in mixed-nationality rosters (IATA, 2020). Such improvements are not only anecdotal; they are reflected in crew resource management audits, LOSA observations, and safety performance dashboards.

6.4.3 Cross-Industry Benchmarking

While aviation presents a unique operational environment, the challenges of multicultural teamwork are shared by other high-stakes domains. In healthcare, for instance, surgical teams operating across cultural boundaries have documented improved patient outcomes when CQ training is integrated into team briefings and simulations. In the maritime industry, where international crews are standard, CQ has been linked to reduced accident rates and improved bridge resource management. These cross-sector insights support the validity of CQ-informed CRM beyond aviation alone. They offer transferable models and comparative lessons, demonstrating that the principles of culturally intelligent teamwork hold across all domains where risk, time pressure, and interpersonal coordination intersect.

6.4.4 Lessons Learned and Best Practices

From these case vignettes, several best practices emerge:

- **Proactive cultural briefing** during pre-flight preparations sets a tone of mutual respect and awareness.
- **Simulated cultural conflict scenarios** allow for risk-free rehearsal of high-stakes interactions.
- **Debriefs focused on cultural dynamics** transform surface-level learning into deep behavioral change.
- **CQ assessments embedded in training pipelines** ensure consistency, accountability, and measurable growth.

Above all, these practices affirm that CQ is not a soft skill, but a safety-critical competency. When supported by organizational commitment and instructional expertise, it becomes an integral pillar of modern CRM – essential for the safe navigation of both cultural and operational turbulence.

6.5 ASSESSMENT AND FEEDBACK MECHANISMS

Dimitrios Ziakkas, Konstantinos Pechlivanis and Hans C. Natakusuma

Assessing the effectiveness of Cultural Intelligence (CQ) within CRM training presents unique challenges. Unlike technical skills or checklist adherence, CQ competencies involve perception, attitude, and behavior across variable cultural contexts. As such, reliable assessment demands a combination of observational metrics, structured feedback, and participant self-reflection.

One widely used foundation for assessment tools is the **Cultural Intelligence Scale (CQS)**, which provides validated items to measure metacognitive, cognitive, motivational, and behavioral CQ (Ang et al., 2007). Adapted versions of this scale can be embedded within CRM simulation debriefs or self-assessment logs. In operational contexts, tailored observation checklists – aligned with EBT/ CBTA principles – can be developed to measure behaviors such as cultural adaptability, responsiveness to cross-cultural input, and style-flexible communication.

For example, during simulator sessions, instructors can use CQ-CRM observation templates that include behaviorally anchored ratings for indicators like:

- Clarifying ambiguous phrases in cross-cultural contexts (Cognitive/ Behavioral CQ)
- Demonstrating sustained engagement with culturally different crewmembers (Motivational CQ)
- Reflecting on how culture may have shaped decision-making (Metacognitive CQ)

These assessments should be applied during **both high-fidelity simulations and post-flight AARs**, allowing instructors to capture CQ in dynamic and unpredictable situations – much like real flight operations.

6.5.1 Validation of Assessment Instruments

Developing a CQ-focused evaluation framework necessitates **empirical validation** to ensure that instruments are both reliable and meaningful. Validation steps include:

- **Expert panel review** of behavior markers by cross-cultural psychologists and aviation SMEs.
- **Pilot testing** in simulator environments with diverse crew compositions.
- **Statistical analysis** of inter-rater reliability to ensure consistency in scoring among evaluators.

Several airlines globally have worked in collaboration with academic partners to develop proprietary CQ evaluation rubrics, which are now integrated into instructor standardization programs and training audits (Livermore, 2024).

6.5.2 Observer Training

Reliable CQ evaluation also depends on the observer – not just the instrument. CRM trainers must be trained not only to recognize cultural variation, but to interpret it **contextually and ethically**. Cultural norms influence behavior in ways that may appear deviant through a monocultural lens. For instance, a reserved response may be interpreted as disengagement, when in fact it reflects cultural values of modesty or respect for authority. Observer training should include:

- **Case-based calibration sessions**, where trainers score video recordings of intercultural simulations and compare results.
- **Bias-awareness modules**, helping observers recognize their own cultural assumptions.
- **CQ theory application**, reinforcing the link between observed behaviors and the theoretical dimensions they represent.

The goal is to ensure **inter-rater reliability**, fairness, and sensitivity in all performance evaluations. This is especially vital when assessments influence promotion, retention, or operational readiness.

6.5.3 Feedback Delivery Methods

The way CQ-related feedback is delivered can greatly affect its acceptance and utility. Instructors must create feedback environments that emphasize learning over judgment, particularly when cultural identity is involved. Constructive, culturally sensitive feedback may follow a model such as:

1. **Observation**: "I noticed in that scenario you maintained a low verbal profile when the situation escalated."

2. **Interpretation**: "This may be influenced by cultural norms around deference or group harmony."
3. **Impact**: "In this case, it may have limited the assertiveness needed for safety-critical input."
4. **Recommendation**: "How might you adapt in a similar situation with a different crew composition?"

This model assures reflection, supports behavioral growth, and avoids essentializing or stereotyping. Facilitators should encourage dialogue and shared problem-solving, transforming feedback into a collaborative tool for development.

6.5.4 Continuous Improvement

Like CRM itself, CQ training is not a one-off event – it is a continual process embedded in the organization's culture of learning. For CQ to remain effective and relevant, **ongoing evaluation, revision, and enhancement** of training content and methods is necessary – in accordance with the four levels of Kirkpatrick theory.

6.5.4.1 CQ Competence Tracking

Organizations can implement systems for longitudinal tracking of CQ development. This may include:

- Logging CQ performance data across multiple training events.
- Incorporating feedback from line operations, LOSA observations, or safety reports.
- Using anonymized SMS data to identify trends in intercultural miscommunication or decision-making friction.

Over time, these insights inform both individual growth and organizational learning.

6.5.4.2 Curriculum Update Processes

A structured review cycle should be built into CRM program management, allowing trainers to adjust modules based on:

- **New operational challenges**, such as evolving geopolitical crew compositions.
- **Regulatory changes**, such as ICAO or EASA updates to training standards.
- **Participant feedback**, gathered from surveys and debrief sessions.

This dynamic process ensures alignment with **industry expectations and operational needs – feedback**.

6.5.4.3 Refresher and Reinforcement Training

Periodic refresher modules sustain CQ competencies and prevent drift. These can take the form of:

- **Quarterly digital learning packages**, focused on specific CQ dilemmas.
- **Annual workshops**, engaging instructors and frontline staff in collaborative learning.
- **Scenario recycles**, revisiting past intercultural challenges with updated learning goals.

Each reinforcement module should reinforce CQ as a lived, practiced skill – not an abstract theory.

6.5.4.4 Feedback Loops and Quality Assurance

Finally, organizations should close the loop by feeding evaluation data back into curriculum design. This quality assurance cycle mirrors established EBT systems and supports compliance with standards such as EASA Part-ORO and FAA AC 120-51E. It also supports a broader **culture of inclusion and safety**, where cultural diversity is viewed not as a challenge to overcome, but as a resource to cultivate.

6.6 CONCLUSION

Dimitrios Ziakkas

The integration of Cultural Intelligence into CRM training is a timely response to the complex, interconnected nature of today's flight operations. As aviation ecosystem actors become more multinational, traditional models of CRM that overlook cultural variables prove insufficient. Embedding CQ within CRM equips crew members not only with interpersonal awareness but also with operational resilience. Through carefully designed training modules, simulation exercises, reflective debriefings, and robust assessment mechanisms, aviation organizations can foster safer and more cohesive crew dynamics. Ultimately, the culturally intelligent flight deck is not merely an aspiration; it is an operational imperative aligned with global safety standards and human-centered aviation.

REFERENCES

Ang, S., & Van Dyne, L. (2008). Conceptualisation of cultural intelligence: Definition, distinctiveness, and nomological network. In S. Ang & L. Van Dyne (Eds.), *Handbook of cultural intelligence* (pp. 3–15). Routledge.

Ang, S., Van Dyne, L., Koh, C., Ng, K. Y., Templer, K. J., Tay, C., & Chandrasekar, N. A. (2007). Cultural intelligence: Its measurement and effects on cultural judgment and decision making, cultural adaptation and task performance. *Management and Organization Review*, 3(3), 335–371.

EASA. (2021). Guidelines for Competency-Based Training and Assessment (CBTA). https://www.easa.europa.eu/en/newsroom-and-events/news/competency-based-training-and-assessment

Flin, R., O'Connor, P., & Crichton, M. (2013). *Safety at the sharp end: A guide to non-technical skills*. CRC Press.

Helmreich, R. L., & Merritt, A. C. (2000). *Culture at work in aviation and medicine: National, organizational and professional influences*. Ashgate Publishing.

IATA. (2020). *Guidance on evidence-based training implementation*. International Air Transport Association.

ICAO. (2013). Manual of Evidence Based Training. Doc 9995. AN/497.

Knowles, M. S., Holton, E. F., & Swanson, R. A. (2015). *The adult learner: The definitive classic in adult education and human resource development* (8th ed.). Routledge.

Livermore, D. A. (2024). *Leading with cultural intelligence: The real secret to success* (3rd ed.). AMACOM.

Salas, E., Tannenbaum, S. I., Kraiger, K., & Smith-Jentsch, K. A. (2020). The science of training and development in organizations: What matters in practice. *Psychological Science in the Public Interest*, 13(2), 74–101.

Part III

Accident Analysis Through a Cultural Lens

7 Methodologies for Cultural Accident Analysis

7.1 INTRODUCTION

Dimitrios Ziakkas

Aviation accident investigation has long been rooted in technical and procedural evaluations. Traditionally, investigative frameworks (e.g., Human Factors Analysis and Classification System (HFACS) or the Swiss Cheese Model (Reason, 1990)) have provided structures for isolating individual, organizational, or systemic breakdowns. However, the increasing complexity of globalized aviation has introduced new dimensions to crew dynamics – namely, the profound impact of culture. Despite the multicultural nature of modern flight decks, cultural influences remain inconsistently analyzed during formal safety investigations. While CRM has evolved to include some aspects of intercultural competence, the application of CQ to accident analysis remains underexplored. Multinational crews became the operational norm. Failure to consider how CQ-related competencies (or their absence) – as discussed in Chapter 4 – affect flight safety represents a significant investigative gap. While individual case studies (e.g., Tenerife 1977; Avianca Flight 052) have hinted at cultural miscommunication, a systematic and codified approach to analyzing cultural contributors is lacking. The proposed approach introduces a methodology for integrating CQ into accident investigations. Drawing upon the ICAO ADDIE model and regulatory guidelines from EASA, FAA, ICAO, and UK AAIB, we propose an interdisciplinary methodology that also accommodates emerging technologies such as AI.

This chapter argues for a more human-centered, data-informed framework capable of interpreting how social attitudes, national cultures, and interpersonal behaviors intersect with operational decisions. By enhancing existing methodologies with CQ-informed codes and culturally sensitive data collection techniques, investigators can better identify the latent conditions that lead to accidents. This approach does not replace current investigative protocols but augments them, positioning cultural competence as a crucial factor in understanding human performance in high-stakes environments.

DOI: 10.1201/9781003672029-10

7.2 INTEGRATIVE ACCIDENT INVESTIGATION FRAMEWORKS

Dimitrios Ziakkas, Anastasios Plioutsias and Thanasis Binis

Traditional accident investigation models, while effective in technical causality and human error classification, often fail to account for cultural dynamics. HFACS, for instance, categorizes human factors contributing to accidents into layers ranging from unsafe acts to organizational influences (Wiegmann & Shappell, 2003). However, the cultural variables embedded within these layers are frequently under-examined or generalized.

In contrast, a culturally informed model systematically integrates the CQ framework within each investigative layer. For instance, a crew's failure to communicate assertively during an emergency may traditionally be coded as a breakdown in CRM. A CQ-informed model would further dissect whether cultural norms about hierarchy or conflict avoidance contributed to the miscommunication (Helmreich & Merritt, 1998). By incorporating CQ, the model captures how cultural variables shape behavior, decision-making, and interpersonal dynamics. This extension is not merely academic. Consider Korean Air Flight 801 (1997), where the first officer's reluctance to question the captain's decision-making was partially attributed to cultural deference. A culturally informed HFACS framework would categorize this behavior not only under communication but also under motivational CQ, identifying a deeper behavioral trend that traditional models may miss.

7.2.1 Data Collection Techniques

Capturing cultural dynamics requires expanding the scope of traditional data collection methods. Standard tools such as Cockpit Voice Recorders (CVRs), Flight Data Recorders (FDRs), and maintenance logs must be complemented by qualitative techniques designed to reveal cultural influences. Three methods are emphasized: structured interviews, behavioral observation, and tailored surveys. Structured interviews should be designed to elicit culturally influenced decision-making processes. For example, questions may focus on how rank affected a crew member's willingness to speak up or whether local customs influenced pre-flight briefings. These interviews should be supported by trained investigators familiar with cultural psychology and the CQ framework. Behavioral observations conducted during simulations or real-time operations can also yield valuable data. Observers can identify verbal and nonverbal indicators of CQ, such as adaptive communication strategies or behavioral rigidity. These observations are most effective when guided by a standardized coding scheme, such as the 20-item CQ taxonomy developed earlier in this book. Tailored surveys administered post-incident or as part of a proactive safety program can quantify CQ dimensions across organizations. These tools should be developed using validated instruments (e.g., the Cultural Intelligence Scale by Ang et al., 2007) and integrated into airline Safety Management Systems (SMS).

7.2.2 Analytical Approaches

Accident investigators must move beyond purely technical or linear causal analysis toward mixed-methods research capable of capturing cultural complexity. The proposed approach combines qualitative content analysis with quantitative modeling to assess the influence of cultural behaviors and social attitudes. Qualitative analysis involves coding interviews and transcripts using the 20-item CQ taxonomy, allowing for the identification of patterns across incidents. This method provides a rich, contextual understanding of how culture shaped crew performance, communication, and risk perception. For example, repeated occurrences of MCQ5 (ignoring cultural context) may suggest a systemic issue in training or organizational culture. Quantitative analysis can include statistical techniques such as correlation analysis, logistic regression, or Bayesian network modeling. These tools enable researchers to assess the predictive power of cultural variables on safety outcomes. For example, investigators may find that low motivational CQ scores correlate strongly with failure to assert during emergencies. In practice, combining qualitative and quantitative methods produces a more comprehensive view. This mixed-methods design is supported by ICAO's ADDIE framework, which emphasizes both needs analysis and evaluative feedback (ICAO, 2022). It ensures that cultural considerations are embedded in both the investigation and the subsequent training redesign.

7.3 PROPOSED FRAMEWORK

Konstantinos Pechlivanis and Dimitrios Ziakkas

The CQ-based lens framework combines recognized accident investigation techniques with standardized CQ indicators. It begins by mapping each CQ factor onto the well-known sequence of events in an occurrence. At the onset, analysts gather standard flight data recorder and cockpit voice recorder transcripts, crew testimonies, and any supplementary logs. The framework then asks investigators to flag evidence of cultural influence that might connect to each of the CQ dimensions.

In tandem, the framework suggests using a coding system that catalogs references to group norms, language usage, or cross-cultural friction. For instance, a remark from a first officer that "I did not challenge the captain because of his seniority" would align with the motivational dimension, since it highlights reluctance to intervene. Similarly, confusion over local phraseology or operational etiquette might be coded under cognitive. Such systematic labeling supports data aggregation, enabling trends to be detected across many events. Once coded, the findings can feed into advanced statistical methods or feed back into operational training programs. The rationale is to position CQ as a supportive angle that illuminates root causes often overlooked.

Developing and implementing this framework requires balancing standard investigatory rigor with an expanded scope of cultural factors. Prior examples in organizational research suggest that once analysts adopt a consistent lens, they detect patterns that previously remained hidden under generic headings like "CRM breakdown." The expected outcomes involve sharper explanations of how crews adapt or fail to adapt under stress. That clarity may then guide airline management to refine training modules. It may also encourage regulators to incorporate cultural intelligence sections in official guidance.

Below is a 20-item coding system aligned with the four core dimensions of CQ: Metacognitive, Cognitive, Motivational, and Behavioral. Each coding item can be used to analyze incident/accident reports, interview transcripts, cockpit voice recorder (CVR) data, or other documents for evidence of cultural intelligence or its absence. The proposed **CQ coding system** offers a structured way to identify, categorize, and analyze cultural intelligence factors in aviation incidents or day-to-day operations. Investigators and researchers can systematically **quantify** cultural contributors to safety outcomes and develop **targeted interventions** to enhance CQ in the aviation community.

The proposed structure is summarized in Table 7.1 and follows the format of:

- **Code** (e.g., MCQ1): A shorthand label for reference.
- **Definition**: A concise explanation of what the code represents.
- **Indicators**: Specific cues or behaviors that investigators should look for.
- **Example**: Illustrative scenario or quote.

The practitioners can utilize the suggested coding system by following five steps:

1. **Data Extraction**
   a. Apply these 20 codes to **transcripts, interviews, incident reports**, and **safety databases** where cultural factors are suspected.
   b. Mark **positive evidence** of CQ (e.g., adapting communication) or **negative markers** (e.g., ignoring cultural context).
2. **Multiple Codes per Incident**
   A single incident might show multiple CQ factors. For instance, a scenario where a pilot misinterprets local norms (CCQ5) and later fails to reflect on it (MCQ5).
3. **Inter-Rater Reliability**
   a. Train investigators on these definitions and examples to ensure **consistent coding** (e.g., using Cohen's Kappa).
   b. Reconcile differences through discussion and consensus.
4. **Statistical Aggregation**
   Once coded, you can **aggregate** frequencies, look for **co-occurrences** (e.g., repeated pairing of MCQ5 and BCQ5), and correlate with **outcome severity** or other contextual factors.
5. Feedback Loop
   Summaries of CQ coding can inform **training** (CRM programs, cultural briefings) or **policy updates** (e.g., reinforcing the need for formal cultural intelligence modules in airline SOPs).

7.4 REGULATORY AND METHODOLOGICAL COMPARISONS

Dimitrios Ziakkas, Thanasis Binis and Anastasios Plioutsias

A comparative review of major regulatory frameworks reveals a fragmented approach to incorporating cultural dimensions into accident investigation. ICAO Annex 13 (Aircraft Accident and Incident Investigation) outlines procedures and structures but

TABLE 7.1
Proposed CQ investigative coding system

Code	Definition	Indicators
MCQ1	**Pre-Interaction Planning**	Evidence of preparing communication strategies, reviewing cultural norms before flights/meetings.
MCQ2	**Real-Time Monitoring**	Actively monitoring and adjusting one's approach during cross-cultural interactions.
MCQ3	**Post-Interaction Reflection**	Reflecting on cross-cultural exchanges to identify successes or improvements.
MCQ4	**Adaptive Strategy**	Demonstrating or lacking a conscious plan to handle cultural differences under stress/emergency.
MCQ5	**Ignoring Cultural Context** (Negative Marker)	Clear lack of awareness or disregard for cultural nuances.
CCQ1	**Awareness of Hierarchy Norms**	Recognizing or misunderstanding different power distance relationships (e.g., deference to rank).
CCQ2	**Understanding Communication Styles**	Knowledge of direct vs. indirect communication preferences across cultures.
CCQ3	**Knowledge of Nonverbal Cues**	Awareness that gestures, eye contact, or physical distance vary culturally.
CCQ4	**Local Procedures & Etiquette**	Familiarity with region-specific or airline-specific cultural/operational norms.
CCQ5	**Cultural Misconceptions** (Negative Marker)	Holding incorrect assumptions or stereotypes about another culture's norms.
MQC1	**Willingness to Engage**	Actively seeks interaction with culturally diverse peers or ground staff.
MQC2	**Persistence & Confidence**	Continuing to communicate/resolve issues despite language barriers or cultural friction.
MQC3	**Positive Attitude Toward Learning**	Enthusiasm for cross-cultural training, reading, or debriefs.
MQC4	**Seeking Feedback**	Actively looking for input on whether communication was clear in multicultural settings.
MQC5	**Avoidance or Apathy** (Negative Marker)	Reluctance to collaborate or learn about other cultures or showing indifference.
BCQ1	**Verbal Adaptation**	Adjusting speech style (tone, formality, clarity) based on cultural expectations.
BCQ2	**Nonverbal Adaptation**	Modifying gestures, personal space, or facial expressions to suit cultural norms.
BCQ3	**Context-Appropriate Feedback**	Delivering or receiving feedback in a manner consistent with the cultural context (direct vs. indirect).
BCQ4	**Role Flexibility**	Changing typical cockpit roles (e.g., being more collaborative vs. hierarchical) to accommodate culture.
BCQ5	**Rigid or Inappropriate Behavior** (Negative Marker)	Refusal or inability to modify one's behavior, causing communication breakdown.

stops short of requiring explicit cultural analysis. Similarly, FAA Order 8020.11D and EASA Regulation (EU) No 996/2010 emphasize the collection and analysis of human performance data, yet cultural intelligence is not operationalized within these mandates. The United Kingdom's Air Accidents Investigation Branch (AAIB), although progressive in human factors reporting, also lacks a codified structure for cultural variables. Most agencies address culture tangentially, often through CRM evaluations, rather than integrating CQ as a core investigative lens.

This divergence necessitates a methodological bridge. One such opportunity is the ICAO ADDIE model (Analysis, Design, Development, Implementation, Evaluation), which underpins training system design. Applying ADDIE to accident analysis allows for structured inclusion of cultural data:

- **Analysis**: Identify cultural dynamics involved in the incident (e.g., language barriers, power distance).
- **Design**: Outline the investigation scope to include CQ-based dimensions.
- **Development**: Create culturally attuned interview questions and coding tools.
- **Implementation**: Conduct the investigation using mixed methods informed by CQ theory.
- **Evaluation**: Apply Kirkpatrick's model to assess learning and behavioral changes post-investigation.

Kirkpatrick's Four Levels of Evaluation provide a complementary framework for post-incident learning:

- **Reaction** – Were stakeholders receptive to the inclusion of cultural elements in the investigation?
- **Learning** – Did investigators gain a deeper understanding of CQ principles?
- **Behavior** – Were there observable changes in training or SOPs that reflect cultural insights?
- **Results** – Did the changes lead to measurable safety improvements?

Using ADDIE and Kirkpatrick in tandem ensures that cultural intelligence is not only integrated during analysis but also embedded in the learning ecosystem that follows.

7.5 IMPLEMENTATION OF AI AND EMERGING TECHNOLOGIES

Dimitrios Ziakkas, Konstantina Ioannidi and Ian Roth

Artificial Intelligence (AI) has become instrumental in augmenting aviation safety. In the context of culturally informed accident investigations, AI can enhance both data collection and pattern recognition. Natural Language Processing (NLP) tools, for example, can process vast amounts of cockpit transcripts, maintenance records, and interview data to identify linguistic cues associated with cultural variables. Machine learning algorithms can be trained to detect markers aligned with CQ dimensions. For example, recurrent patterns of deferential language might flag low motivational

CQ or high power distance environments. These insights can guide investigators toward deeper qualitative probes. Moreover, AI-driven dashboards can visualize CQ-coded data across incidents, revealing systemic cultural vulnerabilities. Finally, AI-enhanced Safety Management Systems (SMS) can integrate CQ variables as new safety indicators, triggering alerts or training interventions when cultural risks emerge. However, AI applications must be transparent, ethical, and supplemented by human oversight to prevent overreliance or misinterpretation.

7.6 CONCLUSION

Dimitrios Ziakkas and Anastasios Plioutsias

As aviation evolves within a complex, multicultural framework, traditional accident investigation methods must adapt. This chapter has presented a culturally enriched methodology anchored in the Cultural Intelligence (CQ) framework and supported by the ICAO ADDIE model, regulatory insights, and AI-based technologies. By incorporating metacognitive, cognitive, motivational, and behavioral indicators into the investigative process, safety professionals can uncover latent cultural factors that influence human performance.

The proposed coding system offers a practical, structured approach to capturing and analyzing cultural data across incidents. When combined with qualitative interviews, quantitative analysis, and machine learning tools, the framework enables more comprehensive safety narratives.

Regulatory harmonization remains a challenge, but integrating cultural considerations into Annex 13 procedures and corresponding FAA/EASA guidance would mark a significant step forward. Meanwhile, continuous feedback loops and Kirkpatrick-style evaluations can ensure that insights from culturally informed investigations translate into meaningful safety improvements.

As aviation continues to embrace diversity in its workforce and operational landscapes, culturally intelligent investigation is not a luxury – it is a necessity. The methodologies presented here lay a foundation for future research, training innovation, and operational resilience.

REFERENCES

Ang, S., Van Dyne, L., Koh, C., Ng, K. Y., Templer, K. J., Tay, C., & Chandrasekar, N. A. (2007). Cultural intelligence: Its measurement and effects on cultural judgment and decision making, cultural adaptation and task performance. *Management and Organization Review*, 3(3), 335–371.

Helmreich, R. L., & Merritt, A. C. (1998). *Culture at work in aviation and medicine: National, organizational and professional influences*. Routledge.

ICAO. (2022). *Doc 9941 – Manual on evidence-based training*. International Civil Aviation Organization.

Reason, J. (1990). *Human error*. Cambridge University Press.

Wiegmann, D. A., & Shappell, S. A. (2003). *A human error approach to aviation accident analysis: The human factors analysis and classification system*. Ashgate.

8 An Evidence-Based Approach

8.1 INTRODUCTION

Dimitrios Ziakkas

Cultural influence in aviation accidents necessitates a thorough examination of high-impact events in which cultural factors have either directly caused or aggravated system failures. These case studies illustrate how miscommunication, hierarchical rigidity, and culturally influenced decision-making patterns can undermine safety margins even in the most technically competent flight environments. A widely discussed example is the crash of Korean Air Flight 801 in Guam in 1997. Safety investigators partly attributed the accident to cultural factors (i.e., the influence of high power distance and deference to authority within Korean cockpit culture (Helmreich & Merritt, 2001)). Subsequent CRM reforms in South Korea were specifically cultural in nature, focusing on diminishing cockpit hierarchy and reinforcing assertiveness across ranks. An additional case is the 1990 Avianca Flight 52 crash near New York, caused by fuel exhaustion following a series of miscommunications between the Colombian flight crew and American Air Traffic Control (ATC). The outcome was tragic – an unequivocal instance demonstrating how cultural expectations influence verbal behavior and how cognitive dissonances can result in catastrophic consequences (Orasanu & Fischer, 1997).

8.2 RECURRENT CULTURAL PITFALLS IN AVIATION SAFETY

Anastasios Plioutsias, Odysseas Papaefthymiou and Yolanda Vaiopoulou

Understanding the persistent cultural pitfalls that impact aviation safety requires more than cataloging error codes or procedural lapses. It demands a human-centered narrative that reflects the psychological, cultural, and communicative dimensions of decision-making under stress. The analysis of relevant past accidents offers critical insight into the latent cultural dynamics that often remain obscured in technical reports. Each case study accident reveals how divergent assumptions, shaped by cultural norms around authority, expression, and teamwork, can silently erode safety margins even within technically competent crews.

Rather than treat these as isolated human errors, this section repositions them as reflections of embedded cultural scripts – scripts that shape how people perceive responsibility, communicate urgency, and negotiate authority. These dynamics, often invisible in the moment, become tragically clear in hindsight. The challenge for aviation professionals is to cultivate the metacognitive awareness necessary to identify and disrupt these scripts before they compromise safety.

DOI: 10.1201/9781003672029-11

8.2.1 Cultural Stress Points and CQ Markers

In reconstructing the narratives of aviation accidents, certain decision points emerge that mark the confluence of technical operations and cultural friction. These "cultural stress points" represent moments when conflicting expectations – whether about how to speak, when to defer, or who should decide – become operational liabilities.

For instance, in the Avianca 52 incident, the captain's failure to explicitly declare a fuel emergency was not merely a linguistic error but a culturally influenced communication gap. The Colombian crew operated within a high-context culture that relies heavily on implied meaning and shared understanding. American air traffic controllers, grounded in a low-context, direct communication model, missed these subtle cues. This misalignment, exacerbated by time pressure, led to fatal consequences (Orasanu & Fischer, 1997).

Similar dynamics played out in Korean Air 801, where hierarchical deference prevented the first officer and flight engineer from asserting themselves when it became clear the captain was continuing a hazardous descent. These cultural inhibitions were not cognitive failures but socially reinforced norms about authority and obedience – norms that, unaddressed, contribute to silence when voice is most needed (Helmreich & Merritt, 2001).

Within these critical junctures, markers of Cultural Intelligence (CQ) become evident. A **metacognitive CQ** response might have prompted the first officer to pause and reflect: "Is my deference helping or harming?" A **behavioral CQ** approach could involve shifting tone or employing strict ICAO phraseology to ensure clarity and assertiveness. These missed micro-interventions underscore the value of embedding CQ within training, not as an abstract theory but as an operational skill set (Ang & Van Dyne, 2008).

8.2.2 Themes of Cultural Breakdown in CRM

Across multiple accident investigations and CRM evaluations, recurring cultural themes have been identified that directly undermine flight crew effectiveness in multicultural settings, as follows:

- **Hierarchical Deference**
  In high power distance cultures, authority figures are seldom challenged, even when subordinate crew members detect anomalies. This was evident in Korean Air 801, where the steep authority gradient silenced dissenting input during the final approach phase. In such contexts, hierarchy becomes a psychological barrier to safety rather than a facilitator of order (Hofstede, 2001).
- **Indirect Communication**
  In high-context cultures, meaning is conveyed implicitly, relying on tone, shared history, and unspoken cues. This stands in contrast to the aviation domain's reliance on low-context, standardized phraseology. In Avianca 52, the assumption that urgency was self-evident through intonation proved fatal when ATC failed to respond to implied distress (Orasanu & Fischer, 1997).

- **Conflict Avoidance and Emotional Restraint**
  Multicultural crews often avoid direct confrontation, particularly when national or professional cultures discourage open disagreement. This creates an environment where concerns are internally suppressed rather than collaboratively addressed, especially under time pressure or when facing senior colleagues.
- **Assumption Bias**
  Crews operating across cultural boundaries may project their own norms onto others – assuming, for example, that what they perceive as respectful communication will be interpreted similarly by others. These misaligned assumptions breed miscommunication and delay decision-making in high-stakes scenarios

(Ang & Van Dyne, 2008; Edmondson, 2019).

8.3 PAKISTAN INTERNATIONAL AIRLINES (PIA) FLIGHT 8303 CASE STUDY

Dimitrios Ziakkas, Anastasios Plioutsias and Konstantinos Pechlivanis

The accident of Pakistan International Airlines Flight 8303 on May 22, 2020, remains a stark reminder of how cultural, organizational, and communicative dimensions can quietly converge to erode flight safety. While the primary accident investigation conducted by Pakistan's Aircraft Accident Investigation Board (AAIB) emphasized procedural violations, overconfidence, and lack of situational awareness, a deeper analysis suggests a potent interplay of cultural intelligence deficiencies and systemic blind spots. Reframing this accident through a Cultural Intelligence (CQ) lens allows for a more textured understanding of the decisions and dynamics that culminated in disaster.

The aircraft – a PIA Airbus A320 – was on final approach to Karachi's Jinnah International Airport. Deviating significantly from standard operating procedures (SOPs), the aircraft approached at excessive altitude and speed. Compounding these technical discrepancies, the flight crew attempted a gear-up landing, scraping the engines along the runway. A subsequent go-around was attempted, but the severely damaged engines failed, leading to a catastrophic crash into a residential neighborhood. Ninety-seven of the 99 people onboard were killed, as were several on the ground.

While technical failures and procedural lapses were obvious contributors, the incident also revealed subtler, culturally rooted dynamics. These included steep cockpit authority gradients, communication inhibition, organizational complacency, and cultural deference – each of which aligns with well-documented constructs in cross-cultural psychology and organizational behavior literature (Helmreich & Merritt, 2001; Gelfand et al., 2007; Livermore, 2024). Through the CQ model's four domains – metacognitive, cognitive, motivational, and behavioral – we gain a structured yet human-centric insight into how these cultural variables may have played out during the final critical moments.

8.3.1 Reassessing Metacognitive CQ: Situational Monitoring and Adaptive Thinking

Metacognitive cultural intelligence refers to the degree to which individuals actively monitor, interpret, and adjust their cultural assumptions during interactions (Ang & Van Dyne, 2008). In the case of PK8303, there appears to have been a significant shortfall in anticipatory awareness. The crew did not appear to conduct a comprehensive briefing tailored to the local airspace complexity, nor did they revise their approach strategy in response to ATC alerts about their unstable trajectory. These missed opportunities for cognitive recalibration highlight the absence of adaptive metacognitive behavior – one that would have demanded a critical reassessment of the deteriorating flight profile and the social dynamics in the cockpit.

The absence of timely intervention by the First Officer (FO) when it became apparent that the aircraft was too high and fast reflects a failure in shared mental modeling – a core aspect of CRM that metacognitive CQ aims to strengthen. In hierarchical environments, such as those commonly found in South Asian cockpit cultures, junior officers may struggle to challenge authority figures even when safety is compromised. This aligns with findings from Hofstede's cultural dimensions, particularly the high power distance observed in Pakistani organizational settings (Hofstede Insights, 2023). In such cultures, the captain's authority is rarely questioned, and this deference can become a systemic risk factor during abnormal operations.

8.3.2 Cognitive CQ: Authority Norms and Communication Perceptions

Cognitive CQ pertains to knowledge of norms, practices, and conventions in different cultures, including professional subcultures such as aviation. The FO's reluctance to voice dissent despite a clearly unstable approach may stem from ingrained hierarchical norms and an internalized assumption that challenging a superior, especially mid-flight, could be interpreted as disrespectful. Furthermore, there may have been an underappreciation of how communication expectations differ between professional domains. For instance, while ATC operates within a low-context, explicit communication paradigm, pilots from high-context cultures might rely on subtle cues or non-verbal affirmations, leading to dangerous misalignments (Orasanu & Fischer, 1997).

This dissonance was evident in the interactions between the cockpit and Karachi ATC. Despite multiple advisories from ATC warning that the aircraft was too high, these were acknowledged without an effective operational response. The failure to act on such input reflects a breakdown in information processing, suggestive not only of technical lapses but also of a cognitive rigidity – an unwillingness or inability to reevaluate the situation in real time. In environments where organizational bravado or past successes normalize deviance, such cognitive blind spots may be further reinforced.

8.3.3 Motivational CQ: Engagement, Confidence, and Risk Perception

Motivational CQ involves the drive to direct attention and energy toward cross-cultural situations. In the context of PK8303, motivational lapses may be inferred from the crew's failure to interrupt a failing plan and consider alternatives such as an early go-around. The desire to "press on" despite accumulating anomalies – unstable approach, gear-up touchdown, degraded engine condition – may reveal not only individual overconfidence but also a culturally reinforced sense of duty or invincibility. When crews operate in environments where deference and emotional restraint are valorized, motivational CQ can suffer. Personnel may lack the psychological safety to escalate concerns, a phenomenon long associated with unsafe team environments (Edmondson, 2019).

Furthermore, the cultural norm of perseverance in adversity – while admirable in many contexts – may become maladaptive when it overrides operational judgment. The persistence to land under deteriorating conditions may thus reflect a motivation not to adapt but to demonstrate control or fulfill perceived expectations. This echoes the concept of "goal fixation" identified in multiple accident analyses (ICAO, 2013), where adherence to a plan becomes pathological, even suicidal, in high-stakes environments.

8.3.4 Behavioral CQ: Inflexible Execution and Suppressed Dissent

Behavioral CQ relates to one's ability to exhibit appropriate verbal and non-verbal actions when interacting across cultures. In high-pressure aviation environments, this includes not only how commands are delivered but also whether opportunities for feedback are created and respected. The CVR transcript, though not fully released, suggests that communication remained formal and unidirectional, with limited collaborative dialogue or verbal check-ins. The captain's control of the cockpit, both procedurally and interpersonally, appears to have gone unchallenged, reinforcing a rigid behavioral script.

Behaviorally intelligent leaders often employ strategies to democratize communication, such as explicitly inviting feedback or slowing the decision-making tempo to allow reflection (Livermore, 2024). The lack of such behavioral adjustments in the cockpit may have left the FO feeling disempowered, thus contributing to a collapse in team coordination. This behavioral rigidity – essentially, an inability to "read the room" in a culturally sensitive way – has also been noted in other CRM failures, such as those observed in the Tenerife disaster or Avianca 52 (Helmreich & Merritt, 2001).

8.3.5 Organizational Culture and Regulatory Interfaces

The organizational culture within Pakistan International Airlines at the time had been under scrutiny for years. Reports from ICAO audits and regional aviation bodies consistently highlighted deficiencies in SOP adherence, CRM implementation, and regulatory oversight. Although PIA had formally adopted CRM training, its integration into practice appeared superficial, with limited emphasis on CQ or cross-cultural dimensions. This mirrors findings in the literature that suggest a persistent gap between CRM training content and real-world cockpit behaviors – especially in culturally homogeneous or hierarchical organizational environments (Goncalves, Lemos, & Valadares, 2020).

Moreover, while regulatory frameworks like those of ICAO, EASA, and FAA increasingly promote Evidence-Based Training (EBT) and Competency-Based Training and Assessment (CBTA), their effectiveness depends on cultural adaptation. The application of the ICAO ADDIE framework – Analysis, Design, Development, Implementation, and Evaluation – remains incomplete if the "Analysis" phase fails to consider national and organizational cultural variables. Similarly, Kirkpatrick's four-level training evaluation model (reaction, learning, behavior, results) suggests that observable cockpit behaviors – such as failure to challenge unsafe decisions – signal a breakdown at both the learning and behavioral levels of CRM training (Kirkpatrick & Kirkpatrick, 2016).

8.3.6 A Framework for Cultural Transformation in Safety Systems

PIA Flight 8303 serves not only as a tragic accident but also as a cultural case study. It exemplifies how ostensibly technical failures may be underpinned by deep-seated cultural patterns, psychological inhibition, and organizational norms. Applying a CQ framework allows safety professionals, regulators, and training organizations to move beyond superficial procedural analyses and into the realm of behavioral transformation. For this transformation to be effective, it must be institutionalized – embedded in training curricula, safety audits, performance evaluations, and even incident investigation protocols.

CQ must be viewed not as an optional soft skill but as a core component of operational safety, particularly in a globalized aviation sector where multicultural crews are the norm, not the exception. The future of flight safety lies not just in data monitoring and automation but in cultivating human adaptability, intercultural empathy, and the courage to speak – even when culture whispers "stay silent."

8.4 TENERIFE AIRPORT DISASTER CASE STUDY

Anastasios Plioutsias, Konstantinos Pechlivanis and Dimitrios Ziakkas

The Tenerife airport disaster of March 27, 1977, remains the deadliest aviation accident in history, claiming 583 lives when two Boeing 747s collided on the runway at Los Rodeos Airport (now Tenerife North Airport) in the Canary Islands. Beyond the devastating loss of life, the accident stands as a pivotal moment in aviation history that ultimately led to the birth of Crew Resource Management (CRM).

The KLM and Pan Am aircraft had both been diverted to Los Rodeos due to a terrorist incident at Gran Canaria Airport. The sudden influx of traffic at a relatively small airport created considerable congestion on the ground, resulting in parking bottlenecks, long delays, and heightened tension among flight crews. Poor visibility from a rolling fog further complicated the operation. Against this tense backdrop, the KLM flight began its take-off roll while the Pan Am aircraft was still taxiing on the same runway.

At the center of the accident lies a critical breakdown in communication between the KLM captain, his crew, and air traffic control (ATC). The KLM captain, a highly respected senior training captain, had demonstrated a strong, assertive style, which likely influenced the first officer's reluctance to question or delay the take-off clearance. When the captain announced, "We're going," the first officer replied with an ambiguous statement that sounded like confirmation, though ATC had not yet given explicit take-off clearance. The controller's use of the phrase "OK, stand by for take-off, I will call you" added further ambiguity.

Analyzing this incident through the lens of Cultural Intelligence (CQ) and CRM reveals deep-rooted issues related to power distance and communication style. Hofstede's (2001) cultural dimensions identify the Netherlands – home of the KLM crew – as relatively low in power distance, suggesting a culture where subordinates feel more empowered to challenge authority. However, the aviation context – and especially the presence of a senior training captain – appears to have overridden these general cultural traits. In contrast, Pan Am's American crew exhibited a more egalitarian cockpit dynamic, with greater openness in their intra-crew dialogue and a more assertive communication style toward ATC. Yet, they too failed to clarify their position strongly enough to prevent the disaster.

This tragic event illustrates how situational hierarchy, role expectations, and cognitive framing can distort otherwise standard communications, especially under stress and ambiguity. In a setting where time pressure, environmental constraints, and interpersonal dynamics converge, assumptions often replace explicit dialogue. The KLM crew likely assumed clearance had been granted, and the first officer's hesitance to challenge the captain – a phenomenon now widely recognized in CRM literature – contributed to the irreversible outcome.

Tenerife also exposes the limitations of standard phraseology when intersected with differing linguistic interpretations and non-native English usage. Miscommunications stemmed not only from language barriers but from mismatched expectations regarding conversational conventions. For instance, the KLM captain's use of "we're now at take-off" was not a request for clearance but rather an implied action – a style of communication more common in Dutch or Germanic language structures. ATC's response, lacking a firm denial or a reiteration of "hold position," did not interrupt the developing misunderstanding.

The subsequent investigation and industry response to Tenerife were unprecedented in scope. Although the technical factors were clear – fog, congestion, and communication lapses – the behavioral and cultural aspects became the primary focus of reform. In the years that followed, airlines began integrating CRM training into their programs, initially focused on authority gradients and communication protocols. The accident effectively shifted the focus of aviation safety from purely technical to deeply human.

More recently, this event has been revisited through the lens of Cultural Intelligence. The roles of metacognitive CQ – awareness of one's own assumptions – and behavioral CQ – ability to adapt communicative behavior across contexts – are now seen as essential competencies for flight crews. Had the KLM crew engaged in more reflective dialogue or had the first officer been empowered to question, the outcome may have been averted. Similarly, had ATC adopted more culturally adaptive and assertive communication protocols, the ambiguity might have been clarified.

The Tenerife disaster is also a reminder of the enduring role of psychological safety in the cockpit. Hierarchical command structures, when left unchecked, can suppress dissent and slow critical feedback loops. Modern CRM training emphasizes flattening authority gradients when safety is at stake, encouraging assertiveness regardless of rank or experience. These principles, grounded in cultural and behavioral science, have roots in the painful lessons learned from this accident.

In contemporary terms, Tenerife offers ongoing relevance. Today's multinational crews must navigate not only technical systems but complex interpersonal terrains. Cultural misalignments can still contribute to miscommunication, even when language and standard operating procedures are shared. As the aviation industry continues to diversify, incorporating CQ and cultural adaptability into CRM training is no longer optional – it is essential.

Thus, Tenerife was not merely an accident; it was a watershed moment. It highlighted the intersection of culture, communication, and hierarchy in aviation safety. Its legacy continues in every simulator session, briefing, and debriefing that reinforces the human dimensions of flight.

8.5 SYNTHESIS OF FINDINGS AND IMPLICATIONS FOR FUTURE PRACTICE

Anastasios Plioutsias, Konstantinos Pechlivanis and Dimitrios Ziakkas

The synthesis of preceding case studies reveals an enduring truth in contemporary aviation: technical competence alone is insufficient to ensure operational safety in multicultural settings. As aviation operations become more globalized, the cultural dynamics within the cockpit – and the broader organizational environment – assume a critical role in shaping outcomes under stress. Crew Resource Management (CRM), traditionally focused on communication, situational awareness, leadership, and decision-making, must evolve to explicitly incorporate Cultural Intelligence (CQ) as a central component of training and practice.

8.5.1 Integration with CRM Training: A Path Toward Safer Multicultural Operations

Incidents such as Avianca Flight 52, Korean Air 801, and Air India Express IX1344 consistently illustrate that cultural attitudes toward hierarchy, emotional expression, and deference to authority deeply influence flight crew interactions. These influences are not superficial traits but embedded cognitive and social schemas that affect how individuals perceive roles, manage ambiguity, and respond to high-stakes decisions (Helmreich & Merritt, 2001).

To adapt CRM training to this complex intercultural reality, the incorporation of CQ theory must go beyond conceptual awareness. Practical implementation requires case-based learning grounded in real incidents, scenario-based simulations that replicate intercultural misunderstandings, and reflective debriefing strategies designed to foster metacognitive awareness. Simulated CRM exercises should integrate behavioral indicators of CQ – such as verbal assertiveness in high power distance settings or adaptability in interpreting non-verbal cues – to develop nuanced operational fluency across diverse cockpit configurations.

The goal is not to impose a cultural uniformity but to equip aviation professionals with the ability to recognize, interpret, and adapt to the cultural context of their peers. This approach blends the four dimensions of CQ – metacognitive, cognitive, motivational, and behavioral – with evidence-based pedagogy and reflective learning (Ang & Van Dyne, 2008; Livermore, 2015). When cultural intelligence is actively embedded in CRM training cycles, it enables crews to respond to uncertainty not with hesitation but with culturally attuned coordination.

8.5.2 Recommendations for Policy and Practice: Bridging Safety and Inclusion

To institutionalize these lessons, aviation stakeholders must move beyond isolated training initiatives and embrace cultural intelligence as a policy imperative. The case studies presented in this chapter reveal a shared pattern: where CRM fails, culture is often the invisible variable left unaddressed. Rectifying this gap calls for a multi-tiered approach integrating regulatory reform, structural innovation, and leadership development.

Regulatory authorities such as ICAO, EASA, and the FAA should consider the formal adoption of cultural competence standards akin to CRM mandates. This includes defining observable CQ performance criteria, establishing minimum training hours focused on cultural adaptability, and requiring assessment mechanisms that validate intercultural proficiency in operational settings. Furthermore, CRM curricula should be audited for inclusivity and cross-cultural applicability, ensuring they do not inadvertently privilege one normative communication style over another.

Safety Management Systems (SMS) must also evolve. Cultural risk factors – such as reluctance to speak up, misunderstanding of indirect language, or misaligned authority gradients – should be systematically integrated into hazard identification and reporting structures. By embedding cultural markers into data analytics and trend analysis, organizations can unearth latent vulnerabilities previously obscured by mono-cultural assessment frameworks.

Leadership plays an equally pivotal role. Selection and development of captains and managers must prioritize cultural empathy, reflective judgment, and inclusive communication skills. These traits should not be considered soft add-ons but core competencies essential for managing multicultural crews. Organizations must cultivate leadership pipelines that elevate individuals with demonstrated cultural humility and adaptive thinking.

Accident investigation protocols, too, require recalibration. Whether using HFACS, Bowtie, LOSA, or FRAM models, investigators should be equipped to identify cultural patterns in behavior, communication, and organizational practice. Contextualizing crew interactions within their sociocultural backgrounds can reveal misalignments and blind spots that are otherwise misattributed to procedural errors or technical deficiencies.

Beyond aviation, insights from culturally mature sectors such as maritime operations, energy production, and global healthcare can offer cross-sectoral benchmarks. These industries have developed ergonomics frameworks that explicitly address the cultural dimensions of teamwork and situational judgment (Gelfand et al., 2007). Drawing on these examples can inspire innovation in aviation safety design.

Finally, cultural intelligence must be integrated into the daily rhythm of airline operations. Refresher training, recurrent CRM updates, and structured debriefings should include space for intercultural analysis and dialogue. Flight reviews and after-action reports offer valuable opportunities to reflect on cultural variables and recalibrate team strategies. Airlines should also diversify representation across departments – safety, training, flight ops – not only as a demographic commitment but as an operational necessity.

Building a truly inclusive safety culture is not simply about representation; it is about resilience. In a world where aviation systems routinely traverse cultural boundaries, the capacity to adapt, listen, and collaborate across differences becomes a critical safety asset. Integrating cultural intelligence into CRM, safety policy, and organizational structures ensures that diversity is not a risk factor – but a source of operational strength.

8.6 CONCLUSION

Dimitrios Ziakkas

The investigations and reflections presented throughout this chapter converge on a critical realization: culture is not a peripheral variable in aviation safety – it is often the invisible scaffolding that shapes decisions, interactions, and system outcomes. While technological advancements, procedural standardization, and automation have significantly improved aviation's risk margins, they cannot fully compensate for latent cultural dynamics that remain unacknowledged and unaddressed.

From Tenerife to Karachi, the case studies reveal patterns that resist simplification. Hierarchical cockpit norms, culturally rooted communication habits, and divergent understandings of authority and assertiveness have repeatedly undermined crew coordination and degraded decision-making. These human elements do not emerge in isolation. They are embedded within national, organizational, and professional cultures that persist, sometimes subtly and sometimes overtly, across borders and generations.

Crew Resource Management (CRM) has long been aviation's answer to the human variability challenge. Yet, as shown in these incidents, CRM's effectiveness is bound by the cultural assumptions embedded in its delivery and interpretation. A CRM course designed around low-context communication norms may not resonate with, or sufficiently prepare, a crew raised in high-context cultures. Similarly, standard callouts and assertiveness protocols lose their efficacy if they fail to account for psychological safety, deference patterns, or culturally reinforced emotional restraint.

It is within this complexity that **Cultural Intelligence (CQ)** offers a transformative framework – not as a replacement for technical training or regulatory compliance, but as a necessary expansion of what it means to be a competent aviation professional in the 21st century. The four dimensions of CQ – metacognitive, cognitive, motivational, and behavioral – illuminate where breakdowns occur and how they might be prevented. They also offer a language and structure for bridging cultural divides, allowing crews to operate not only with technical synchronization but with interpersonal harmony.

Moreover, the ICAO **ADDIE model** – Analysis, Design, Development, Implementation, and Evaluation – must now be applied with cultural variability in mind. The "Analysis" phase, in particular, must incorporate organizational and national cultural diagnostics before designing or delivering training. Likewise, **Kirkpatrick's four levels of training evaluation** (Reaction, Learning, Behavior, Results) offer a powerful feedback mechanism that can capture whether CQ principles are merely absorbed or actually translated into adaptive cockpit behaviors.

Finally, this chapter's synthesis demonstrates that culture is not a risk factor to be controlled; it is a dynamic force to be understood and cultivated. When properly integrated into CRM training, safety management systems, and leadership development, cultural intelligence becomes a multiplier of resilience, empathy, and precision.

As aviation continues to evolve – into more automated, more global, and more multicultural forms – its safety culture must also evolve. The integration of CQ into CRM, into policies, and into everyday operations is not merely progressive; it is imperative. In a field where safety is measured not only by systems that prevent failure but by people who recognize, communicate, and adapt under stress, cultural competence is not a luxury – it is a frontline defense.

REFERENCES

Ang, S., & Van Dyne, L. (2008). *Handbook of cultural intelligence: Theory, measurement, and applications*. Routledge.

Edmondson, A. C. (2019). *The fearless organization: Creating psychological safety in the workplace for learning, innovation, and growth*. Wiley.

Gelfand, M. J., Erez, M., & Aycan, Z. (2007). Cross-cultural organizational behavior. *Annual Review of Psychology*, *58*, 479–514. https://doi.org/10.1146/annurev.psych.58.110405.085559

Goncalves, L., Lemos, C., & Valadares, C. (2020). Enhancing CRM through immersive training: A cross-cultural perspective. *Journal of Aviation Psychology*, *30*(2), 145–163.

Helmreich, R. L., & Merritt, A. C. (2001). *Culture at work in aviation and medicine: National, organizational and professional influences*. Routledge.

Hofstede, G. (2001). *Culture's consequences: Comparing values, behaviors, institutions and organizations across nations* (2nd ed.). Sage Publications.

Hofstede Insights. (2023). *Country comparison: Pakistan*. https://www.hofstede-insights.com

International Civil Aviation Organization (ICAO). (2013). *Human factors training manual (Doc 9683)*. ICAO.

Kirkpatrick, D. L., & Kirkpatrick, J. D. (2016). *Evaluating training programs: The four levels* (3rd ed.). Berrett-Koehler.

Livermore, D. A. (2015). *Leading with cultural intelligence: The real secret to success* (2nd ed.). AMACOM.

Livermore, D. A. (2024). *Leading with cultural intelligence: The real secret to success* (3rd ed.). AMACOM.

Orasanu, J., & Fischer, U. (1997). Say it again, Sam! Effective communication strategies to mitigate pilot error. In R. S. Jensen (Ed.), *Proceedings of the Ninth International Symposium on Aviation Psychology* (pp. 697–703). Ohio State University.

Part IV

Strategies, Policy, and Future Directions

9 Designing and Evaluating Cross-Cultural CRM Training Programs

9.1 INTRODUCTION

Dimitrios Ziakkas

In today's rapidly evolving aviation industry, globalization has reshaped the nature of crew collaboration. Multinational teams are no longer a rarity – they are the norm. The cockpit and cabin crew are increasingly composed of individuals from vastly different cultural, linguistic, and operational backgrounds. While this diversity brings strength and flexibility, it also presents unique challenges in communication, leadership, trust-building, and decision-making. In this context, the integration of Cultural Intelligence (CQ) into Crew Resource Management (CRM) training is no longer optional – it is a necessity.

Traditional CRM, grounded in standardized protocols and linear communication models, must evolve to account for the fluidity and complexity of intercultural interactions. Misunderstandings rooted in cultural expectations can subtly undermine team coordination, particularly in high-stakes environments. A culturally intelligent CRM framework, therefore, must build on existing competencies while introducing new ones that allow crews to operate effectively across cultural boundaries.

This chapter provides a comprehensive exploration of how cross-cultural CRM training programs can be effectively designed and evaluated. We propose a framework that draws on the ICAO ADDIE model – Analysis, Design, Development, Implementation, and Evaluation – while leveraging the Kirkpatrick model to assess learning outcomes at four progressive levels. Emphasis is placed on integrating CQ's metacognitive, cognitive, motivational, and behavioral dimensions throughout the CRM lifecycle. Drawing from best practices, real-world case studies, and regulatory expectations, we present both the theoretical foundations and practical mechanisms that inform effective cross-cultural CRM.

9.2 TRAINING PROGRAM DEVELOPMENT

Dimitrios Ziakkas, Dimitra Synodinou and Ioanna Lekea

Well-formulated training objectives are foundational to a successful program. These objectives should reflect operational priorities and regulatory standards while integrating cultural dimensions explicitly. The training objective areas focus on the following aspects:

DOI: 10.1201/9781003672029-13

- Trainees will identify and describe at least three culturally influenced behaviors that affect cockpit communication.
- Trainees will demonstrate the ability to adapt verbal and non-verbal communication styles in simulations with culturally diverse team members.
- Trainees will assess cultural dynamics in real-time scenarios and make inclusive decisions that reflect mutual respect and operational clarity.

These objectives align with ICAO's Doc 9995 (2023) and EBT/CBTA models, promoting evidence-based, observable outcomes.

9.2.1 Curriculum Design: A Modular Framework

Using the ADDIE model, training content is designed to move from theoretical knowledge to applied skills:

Module 1: Foundations of Cultural Intelligence Introduces Hofstede's cultural dimensions (Hofstede, 2001), the GLOBE study findings, and their implications for cockpit behavior. It highlights the influence of cultural worldviews on hierarchy, uncertainty, time orientation, and communication preferences.

Module 2: Cross-Cultural Communication in CRM Focuses on the nuances of high-context vs. low-context communication. Through case discussions and flight deck voice recordings, trainees explore how tone, phrasing, and silence are interpreted differently across cultures. Exercises emphasize the importance of clear, concise language, especially under stress.

Module 3: Cultural Influences on Risk and Decision-Making Draws on real-world incidents (e.g., Avianca Flight 52, Air France 447) to illustrate how cultural assumptions can distort risk perception. Scenario-based training challenges crews to identify biases, ask clarifying questions, and develop inclusive decision models.

Module 4: Building Trust and Managing Conflict Explores the role of empathy, listening, and non-verbal sensitivity in fostering cohesion. Trainees engage in conflict resolution simulations where cultural tension is built into the scenario.

Module 5: Sustaining Cultural Competence Includes strategies for continuous learning, such as using cultural briefing apps, participating in intercultural coaching, and embedding CQ into performance feedback systems.

9.2.2 Embedding CQ Dimensions

Effectively embedding CQ into CRM training requires deliberate alignment of the metacognitive, cognitive, motivational, and behavioral dimensions of CQ into each training component. In detail:

- **Metacognitive CQ**
  Pre-flight discussions are structured to surface assumptions about behavior and hierarchy. Debriefings use "reflective questioning" to encourage self-awareness. Facilitators are trained to guide discussions using the Debriefing with Good Judgment framework (Rudolph et al., 2007).

- **Cognitive CQ**
  Trainees receive culture-specific briefings that go beyond etiquette. These include analysis of error reporting norms, uncertainty tolerance, and views on authority. Case studies highlight how misunderstandings can occur despite shared technical knowledge.
- **Motivational CQ**
  Learners are encouraged to embrace intercultural challenge through peer narratives and recognition mechanisms. Participants hear from senior captains who share personal stories about cultural conflict and growth, fostering intrinsic motivation.
- **Behavioral CQ**
  In high-fidelity simulations, learners practice shifting their speech patterns, body language, and assertiveness based on team composition. Facilitators use behavioral rubrics to track adaptability.

9.3 PROPOSED FRAMEWORK – EVIDENCE-BASED LEARNING AND OPERATIONAL SIMULATIONS

Dimitrios Ziakkas and Hans C. Natakusuma

Embedding Cultural Intelligence (CQ) within Crew Resource Management (CRM) training involves more than conceptual integration; it requires pedagogical intentionality, design sophistication, and alignment with the evolving technological and socio-operational landscape. This section expands upon the four core CQ dimensions – metacognitive, cognitive, motivational, and behavioral – by articulating how each can be intricately embedded in CRM design, reinforced through advanced learning technologies, and assessed using evidence-based instructional practices. Furthermore, the section situates CQ not just as an academic construct, but as a lived, practiced competency that enhances safety, cohesion, and decision-making within multinational crews. The proposed framework focuses on the following areas:

9.3.1 Metacognitive CQ: Enhancing Reflection in Contextual Complexity

Metacognitive CQ refers to the mental processes individuals use to acquire and understand cultural knowledge. It is activated when crew members consciously plan for intercultural interactions, monitor their assumptions, and reflect on the effectiveness of their communication and coordination. In the CRM context, this can be achieved through pre- and post-simulation debriefs where facilitators pose targeted reflective questions such as: "What cultural assumptions influenced your interpretation of the captain's tone?" or "How did your understanding of hierarchy shape your decision to speak up?"

Emerging technologies such as AI-supported debriefing platforms are now being used to prompt these questions based on real-time analysis of performance during simulation. These platforms use natural language processing to flag instances of deference, assertiveness, or silence and connect these behaviors to known cultural patterns. Augmented reality (AR) overlays can further deepen reflection by visually reconstructing flight deck interactions, allowing learners to annotate and reflect on cultural cues they may have missed.

Embedding metacognitive CQ in CRM, therefore, requires more than introspection; it involves equipping learners with structured opportunities and intelligent feedback tools that externalize their thinking processes and illuminate hidden biases.

9.3.2 Cognitive CQ: Knowledge Integration Across Domains

Cognitive CQ captures the knowledge individuals have about cultural norms, practices, and conventions. It extends beyond superficial awareness into domain-specific expertise – such as understanding how different cultures perceive authority, uncertainty, and error reporting.

In advanced CRM programs, cognitive CQ is developed through interdisciplinary content that draws from anthropology, organizational psychology, and applied linguistics. For instance, trainees might explore how a collectivist worldview influences a first officer's reluctance to question a captain's command. This understanding is then mapped onto CRM competencies such as decision-making and leadership.

Digital learning environments using adaptive learning algorithms (e.g., LXPs – Learning Experience Platforms) can tailor cognitive CQ instruction based on the learner's cultural background and experience level. Scenario-based modules automatically adjust case studies to feature cultural dynamics most relevant to the learner's current role or anticipated deployment.

Moreover, cognitive CQ can be reinforced through global crew exchanges and virtual collaborations. AI Platforms, which simulate cross-cultural briefings in real time, allow trainees to negotiate communication and protocol across cultures – moving cognitive CQ from the abstract into the operational.

9.3.3 Motivational CQ: Sustaining Cultural Engagement through Emotional Resonance

Motivational CQ relates to the interest, drive, and confidence to function effectively in culturally diverse settings. Without intrinsic motivation, CQ development remains fragile and superficial. In the training environment, motivational CQ is built through emotionally compelling experiences that validate cultural effort.

Storytelling remains a powerful pedagogical tool in this regard. Senior flight crew members sharing moments of intercultural misunderstanding – and resolution – create psychological safety and normalize cultural mistakes as learning opportunities. CRM training programs have begun to embed these narratives in video form within modules, followed by emotion-check surveys that help learners process their reactions and attitudes toward cultural complexity.

Gamified CRM apps offer an advanced path to motivational development. These platforms provide immediate feedback and recognition for successful intercultural navigation, rewarding persistence, curiosity, and empathy. Badging systems and progression metrics, when aligned with organizational performance reviews, further incentivize cultural engagement.

Crucially, motivational CQ should also be addressed in instructor training. Facilitators must be equipped to recognize motivational dips and redirect resistance through curiosity-based questioning, emotional reframing, and modeling of vulnerability in cultural learning.

9.3.4 Behavioral CQ: Operationalizing Cultural Adaptability

Behavioral CQ refers to the capacity to act appropriately across cultural contexts. In CRM, this includes modulating tone, adjusting assertiveness, synchronizing body language, and adapting communication structures to match cultural expectations.

Training for behavioral CQ must be immersive. High-fidelity flight simulators are increasingly integrating wearable biometric feedback devices that measure stress responses during cross-cultural exchanges. These data – such as elevated heart rate variability during a disagreement – can inform debriefs on how cultural friction may manifest physiologically.

Real-time avatar-based role-plays, supported by AI behavioral engines (e.g., SimPILOT-X), allow trainees to engage with synthetic characters from diverse cultural backgrounds. These avatars respond with culturally accurate behaviors, giving trainees practice in real-time adaptation. For example, if a trainee interrupts a virtual Japanese captain, the avatar may respond with subtle cues of discomfort, prompting the learner to adjust their behavior accordingly.

Additionally, behavioral CQ is reinforced through structured observation. Supervisors and CRM instructors use calibrated rubrics to assess interactions during line operations, offering specific feedback on verbal tone modulation, pacing, turn-taking, and gesture management. This creates a continuous improvement loop that extends beyond the classroom.

9.3.5 Cross-Dimensional Integration: Toward a Systems Perspective

While each CQ dimension provides unique value, the real power emerges when they are integrated across the CRM training journey. A simulation might begin with metacognitive planning (e.g., cultural briefing), draw on cognitive knowledge (e.g., understanding of collectivism), require motivational persistence (e.g., navigating a disagreement), and culminate in behavioral adaptation (e.g., adjusting tone and gestures).

This systems-based view aligns with ICAO's evidence-based training philosophy, which encourages dynamic, contextual, and performance-oriented learning. Using the ADDIE model, CRM programs can be structured to revisit CQ dimensions iteratively:

- **Analysis**: Identify the cultural fault lines in existing safety data.
- **Design**: Map CQ competencies to CRM scenarios.
- **Development**: Build immersive content leveraging XR (Extended Reality), NLP (Natural Language Processing), and AI-driven avatars.

- **Implementation**: Deliver training in modular, scalable formats with continuous facilitator support.
- **Evaluation**: Assess via Kirkpatrick's four levels and real-time performance analytics.

Through this integrated, multi-modal approach, CQ becomes not an abstract theory but a practiced, evaluated, and essential component of flight safety. Embedding CQ into CRM training must be understood not as a discrete initiative, but as a transformation of how we prepare, operate, and learn together in culturally diverse teams. It demands that we recognize culture not as an obstacle, but as a condition of modern aviation – and an opportunity for deeper safety, innovation, and respect. Emerging technologies offer the tools to accelerate this integration, but it is the human commitment – by educators, crew, and organizations – that will sustain it. As aviation moves forward into an era of increasing automation and global interdependence, the human capacity for cultural empathy, reflection, and adaptability will define not only the quality of CRM training but the resilience of the system itself.

9.4 EVALUATION AND FEEDBACK

Dimitrios Ziakkas and Ian Roth

In the context of cross-cultural CRM training, case-based learning and interactive simulations provide an essential bridge between theoretical understanding and operational application. This approach aligns with constructivist learning theory, which posits that learners build knowledge through experience and reflection. Through the exploration of real-world scenarios and immersive simulation, flight crews develop not only cognitive awareness of cultural variables but also the behavioral adaptability required in high-pressure, dynamic environments.

The integration of actual aviation events into CRM training transforms passive learning into active inquiry. For instance, the case of Korean Air Flight 801 has long been cited as an example of the threats posed by excessive deference and rigid authority gradients, which are often influenced by national cultural norms. In this incident, the failure of the first officer to assertively challenge the captain's decision-making contributed to a fatal controlled flight into terrain (NTSB, 1999). Through structured analysis, trainees learn to identify cultural barriers that interfere with crew coordination and situational awareness.

A successful evidence-based learning environment must do more than present historical events; it should guide learners through structured debriefings using reflective frameworks like the "Debriefing with Good Judgment" model (Rudolph et al., 2007). This ensures that the lessons extend beyond blame attribution and foster growth in metacognitive and emotional intelligence dimensions.

9.4.1 Simulations and Cultural Complexity

High-fidelity simulations are particularly powerful in embedding cultural intelligence (CQ) competencies into CRM. Unlike static classroom lectures, simulations replicate the complexity, urgency, and uncertainty of real flight operations. When cultural elements are intentionally embedded – such as simulated communication breakdowns due to linguistic misunderstandings, hierarchical hesitations, or misinterpretations of non-verbal cues – trainees are pushed to actively apply the CQ dimensions in real time.

Emerging technologies now allow for greater sophistication in this domain. Virtual Reality (VR) simulations, for instance, provide immersive experiences where avatars with culturally distinct behaviors interact with crew members. AI-powered role-players can adapt their responses based on the trainee's input, generating a responsive training environment that mimics real-world unpredictability. These tools enable personalized learning pathways while collecting data on trainee behaviors, emotional regulation, and decision timelines.

9.4.2 Emerging Technologies and CRM Training

The introduction of digital twins further enhances CRM training. These systems collect data from real flight operations and simulate culturally diverse interactions in cockpit and cabin environments. By incorporating data on crew backgrounds, training histories, and known cultural conflict triggers, digital twins allow instructors to model and forecast potential breakdowns in multicultural settings.

Recent advancements in AI and machine learning provide tools that analyze trainee interactions during simulation in real time. Speech recognition algorithms can detect hesitations, deference patterns, and interruptions, flagging them as indicators of underlying cultural tension. Similarly, natural language processing (NLP) tools evaluate the clarity, tone, and assertiveness of communication.

During debriefing, AI-generated heat maps visualize areas of strong or weak communication in relation to crew positioning, enabling participants to see how cultural variables influenced their decision-making under pressure. These insights support instructors in delivering precise, evidence-based coaching. Furthermore, integrating these technologies aligns with the ICAO competency-based training (CBTA) paradigm, which prioritizes performance indicators and contextual feedback over time-based metrics.

9.4.3 Scenario Design and Cultural Calibration

Designing culturally rich simulations requires collaboration between subject matter experts, human factors specialists, and cultural psychologists. The scenario design process begins with a detailed task analysis to identify points where cultural misunderstandings are likely to surface. For example, pre-flight briefing rituals, command tone during emergencies, and post-landing coordination are commonly rich in cultural subtext.

Each simulation must be calibrated to reflect not only linguistic differences but also culturally encoded decision preferences. Research from the GLOBE study informs how collectivist vs. individualist tendencies influence behavior. In collectivist cultures, decision-making may prioritize group harmony over procedural exactness.

In contrast, pilots from low power distance backgrounds may question authority more readily, creating potential for conflict if such behaviors are misinterpreted.

Calibration also considers factors such as gender dynamics, generational differences, and religious observances. This holistic approach ensures that simulations reflect the full complexity of cross-cultural flight operations.

9.4.4 Structured Reflection and Learning Transfer

Simulation is most effective when paired with structured reflection. Instructors facilitate post-simulation discussions that explore not only what occurred but why it happened, encouraging deep introspection into cultural biases and adaptive strategies. Reflection prompts are designed to engage all four CQ domains:

- Metacognitive: How did I anticipate cultural challenges?
- Cognitive: What cultural norms were at play?
- Motivational: Was I committed to bridging cultural differences?
- Behavioral: Did I adapt my behavior appropriately?

To reinforce learning transfer, participants are assigned post-simulation tasks such as scenario rewriting, peer coaching, or cultural adaptation journals. These tasks ensure that CQ gains extend beyond the classroom into actual line operations.

Evidence-based simulation training benefits further from blended learning models that combine asynchronous digital modules, instructor-led discussions, and on-the-job application. Online modules can cover foundational CQ concepts and cultural briefings for destination regions, allowing face-to-face time to focus on experiential learning. Mobile learning platforms offer refresher content on cultural dos and don'ts prior to duty, reinforcing learning at the point of need.

Furthermore, the proposed approach follows the four levels of Kirkpatrick's theory (Kirkpatrick and Kirkpatrick, 2006), as follows:

- **Kirkpatrick Level 1 – Reaction**
  Learners complete surveys rating the relevance of content and the psychological safety of the learning environment. Open-ended responses capture emotional reactions and cultural comfort levels.
- **Kirkpatrick Level 2 – Learning**
  CQ assessments such as the CQS (Ang et al., 2007) are administered pre- and post-training. Improvements are tracked and analyzed by subgroup (e.g., regional background, flight experience).
- **Kirkpatrick Level 3 – Behavior**
  Observational assessments during line operations evaluate the application of CQ behaviors. Behavioral markers include speaking up, adapting tone, and inviting input from culturally reticent peers.
- **Kirkpatrick Level 4 – Results**
  Organizations track CRM-related incident rates, pilot reports of communication breakdowns, and anonymous crew climate surveys to determine the training's long-term impact.

9.5 BEST PRACTICES FROM GLOBAL OPERATORS

Dimitrios Ziakkas

The successful implementation of a cross-cultural CRM training program cannot be gauged solely by the strength of its curriculum design or the sophistication of its simulation technology. True effectiveness lies in the seamless and sustainable transfer of acquired cultural competencies from the training environment to daily operations. This section elaborates on the critical phase of training transfer – how knowledge, attitudes, and behaviors shaped during CRM training translate into real-world performance across multicultural flight crews and operational settings. By expanding on transfer mechanisms, workplace integration strategies, and supporting infrastructure, we present a framework that not only sustains learning but embeds it deeply into organizational culture. The following recommendations present a step-by-step approach to best practices from global operators.

- **Understanding the Transfer Gap**
  One of the recurring challenges in aviation training, particularly in the domain of CRM and cultural intelligence, is the situation known as the "transfer gap." Despite well-designed training, learners often struggle to apply new skills under operational stress or within organizational cultures that do not reinforce adaptive behaviors. Research by Baldwin and Ford (1988) identified three key factors influencing training transfer: trainee characteristics (e.g., motivation, self-efficacy), training design (e.g., relevance, practice opportunities), and work environment (e.g., peer and managerial support). Applying these principles to cross-cultural CRM underscores the importance of aligning post-training reinforcement strategies with the values and demands of real-world flight operations.
- **Just-in-Time Support Systems**
  To bridge this gap, just-in-time (JIT) support mechanisms have gained traction in modern CRM programs. These include digital applications that provide instant cultural briefings on specific destinations or crew profiles prior to flight, as well as wearable devices that offer real-time prompts or alerts about culturally sensitive interactions. For instance, a pilot flying with a crew from a high power distance culture might receive a pre-departure alert encouraging explicit invitations for feedback during briefings – a tactic aimed at countering cultural reluctance to speak up. Such interventions are subtle yet impactful, facilitating behavioral change without overwhelming the user.
- **AI-Enhanced Behavioral Feedback**
  The integration of Artificial Intelligence (AI) offers promising avenues for monitoring and guiding the application of CRM competencies in operational settings. Emerging platforms, such as CrewBridge AI or Cognition-Flight, use natural language processing (NLP) to analyze cockpit communications in real time, providing immediate feedback on clarity, tone, and inclusion. These systems can detect when a captain dominates discussions without soliciting input, or when culturally loaded terms may cause

confusion. Over time, such platforms generate personalized development trajectories for crew members, highlighting strengths and areas for growth in intercultural communication and team coordination.

- **Managerial Coaching and Learning Contracts**
  Training transfer is further enhanced when organizational leadership plays an active role in reinforcing culturally intelligent behaviors. One effective method involves the use of structured learning contracts, co-developed by trainees and their supervisors at the end of the training program. These contracts outline specific goals related to cultural competence, such as initiating inclusive debriefings or adjusting verbal commands for non-native English speakers, and set time-bound milestones for their achievement. Managers provide feedback and support, transforming cultural intelligence from an abstract competency into a visible performance metric. This participatory model echoes the principles of transformative learning theory (Mezirow, 1991), where reflection and dialogue catalyze enduring change.
- **Peer-Supported Transfer and Cultural Mentorship**
  Learning communities and peer-to-peer support play an essential role in consolidating training outcomes. Airlines can foster this through cultural mentorship programs, where experienced pilots or cabin crew with high CQ are paired with newer employees navigating multicultural teams for the first time. These mentors provide not only operational guidance but also narrative-based learning through storytelling, a pedagogical method known for deepening empathy and memory retention (Clark & Rossiter, 2008).
- **Simulated Reinforcement and Refresher Modules**
  While initial CRM training may occur in controlled environments, long-term retention requires recurrent exposure to realistic cross-cultural challenges. Refresher modules – delivered quarterly or biannually – should incorporate updated case studies, current geopolitical shifts affecting cultural norms, and new scenarios harvested from recent SMS (Safety Management System) data. Virtual reality (VR) and mixed-reality platforms now make it possible to simulate increasingly complex, time-sensitive interactions that stretch cultural competencies under realistic constraints. For instance, trainees may be placed in a high-tempo turnaround scenario at an unfamiliar airport where local ATC phraseology, differing crew hierarchies, and language barriers converge. Repeated exposure in such scenarios improves pattern recognition and builds intercultural resilience.
- **Measurement of Transfer in Operational Contexts**
  Operationalizing cultural intelligence in the cockpit or cabin requires tools that quantify its behavioral expression. Airlines can deploy behavioral checklists and observation protocols that focus on inclusive language, equitable task distribution, and active listening indicators. These checklists can be embedded in routine line checks, feedback forms, or integrated into pilot self-assessments. Airlines introducing the use of digital observation forms that prompt instructors to assess crew engagement across four CQ dimensions during actual flights. This real-time data helps validate training effectiveness and identifies areas for remedial coaching.

- **Institutionalizing CQ within SOPs and Manuals**
  For long-term cultural transformation, training outcomes must become embedded in the airline's operational DNA. Standard Operating Procedures (SOPs), flight manuals, and cabin service protocols should include language and structures that reflect cultural intelligence principles. This may include standard phrases that reduce ambiguity for non-native speakers, structured briefing templates that invite input from junior crew, or checklist items that prompt cultural awareness in unfamiliar locations. By anchoring CQ in institutional documents, airlines signal its importance not just as a training topic, but as a core component of safe and effective operation.
- **Organizational Learning and Feedback Loops**
  A dynamic training transfer model also requires systematic feedback loops. Post-flight surveys, incident debriefings, and CRM reflection sessions must routinely explore whether cultural dynamics played a role in performance, miscommunication, or decision-making. Trends identified through Safety Management Systems (SMS) and Flight Data Monitoring (FDM) should be cross-referenced with cultural indicators to inform future training content. An airline that notices repeated friction between mixed-nationality crews in specific geographic regions, for instance, can develop targeted microlearning modules or pre-departure briefings tailored to that cultural context.

9.6 CONCLUSION

Dimitrios Ziakkas

Cross-cultural CRM training is not merely a pedagogical innovation – it is a safety imperative. In a world where crews may share an aircraft but not a worldview, understanding and bridging cultural gaps becomes the cornerstone of operational resilience. The integration of Cultural Intelligence into CRM must be deliberate, data-driven, and deeply human. Through the careful application of instructional frameworks like ADDIE and robust evaluation models such as Kirkpatrick's, aviation organizations can craft learning experiences that not only meet regulatory demands but also enhance safety, dignity, and performance across every flight deck.

Effective transfer of cultural intelligence from CRM training to operational performance is not accidental; it is engineered through deliberate design, technological support, and cultural reinforcement. By embracing AI-enhanced tools, immersive simulations, peer mentorship, and managerial accountability, airlines can close the gap between what is learned in the classroom and what is practiced in the sky.

As aviation becomes more automated and globalized, the human element – our capacity to listen, adapt, and respect – remains as vital as ever. Cultural competence, therefore, is not a soft skill but a critical one, essential to aviation's future.

REFERENCES

Ang, S., Van Dyne, L., & Koh, C. (2007). Cultural intelligence: Its measurement and effects on cultural judgment and decision making, cultural adaptation and task performance. *Management and Organization Review*, *3*(3), 335–371.

Baldwin, T. T., & Ford, J. K. (1988). Transfer of training: A review and directions for future research. *Personnel Psychology*, *41*(1), 63–105.

Clark, M. C., & Rossiter, M. (2008). Narrative learning in adulthood. *New Directions for Adult and Continuing Education*, *2008*(119), 61–70.

Hofstede, G. (2001). *Culture's consequences: Comparing values, behaviors, institutions and organizations across nations* (2nd ed.). Sage Publications.

ICAO. (2023). *Manual on evidence-based training* (Doc 9995). https://www.icao.int

Kirkpatrick, D. L., & Kirkpatrick, J. D. (2006). *Evaluating training programs: The four levels* (3rd ed.). Berrett-Koehler.

Mezirow, J. (1991). *Transformative dimensions of adult learning*. Jossey-Bass.

NTSB. (1999). *Aircraft accident report: Korean Air Flight 801*. National Transportation Safety Board.

Rudolph, J. W., Simon, R., Rivard, P., Dufresne, R. L., & Raemer, D. B. (2007). Debriefing with good judgment: Combining rigorous feedback with genuine inquiry. *Anesthesiology Clinics*, *25*(2), 361–376.

10 Policy, Organizational Change, and Future Trends

10.1 INTRODUCTION

Dimitrios Ziakkas

In today's globally networked aviation environment, the notion of leadership extends well beyond operational efficiency and into the cultural and psychological dimensions of team dynamics. As aviation crews become increasingly diverse, airlines operate across varied cultural landscapes, and stakeholders come from multiple sociocultural backgrounds, the need for culturally intelligent leadership becomes essential – not optional. Traditional leadership models, although effective in structured environments, often fall short in addressing the nuanced interpersonal challenges that arise in multicultural flight operations.

CQ (Ang and Van Dyne, 2008) provides a multidimensional framework through which aviation leaders can navigate the complexities of multicultural interactions. When applied to leadership, CQ goes beyond technical or procedural competence – it captures the ability to relate and work effectively across cultures through cognitive, motivational, behavioral, and metacognitive dimensions. The real test of leadership in modern aviation, therefore, is the capacity to apply these dimensions in real-time decision-making, crisis management, team building, and conflict resolution across globalized crews.

Furthermore, this expanded leadership paradigm demands alignment with established safety and training frameworks, such as the ICAO's ADDIE model and the Kirkpatrick Evaluation Model. ICAO's ADDIE (Analysis, Design, Development, Implementation, Evaluation) approach ensures that cultural leadership training is grounded in instructional design best practices, while Kirkpatrick's model provides the evaluative scaffolding needed to assess outcomes at the level of reaction, learning, behavior, and organizational results (Kirkpatrick & Kirkpatrick, 2006). Leaders equipped with high cultural intelligence can foster inclusive environments where diversity is not merely tolerated but actively leveraged to enhance safety and innovation.

In this chapter, we examine the critical intersection of leadership, culture, and safety within aviation, exploring how corporate cultures shape safety norms and CRM strategies, how culturally intelligent leadership can enhance crew effectiveness, and how policy frameworks can institutionalize CQ across organizations. Drawing from best practices in major carriers and regulatory bodies – including ICAO, EASA, FAA, and IATA – we synthesize a holistic strategy for embedding

DOI: 10.1201/9781003672029-14

cultural competence into leadership development programs and CRM systems. In doing so, we argue that culturally intelligent leadership is not just beneficial but foundational for the resilience, safety, and ethical integrity of the global aviation system.

10.2 POLICY IMPLICATIONS FOR AVIATION SAFETY

Dimitrios Ziakkas and Anastasios Plioutsias

In contemporary aviation, global interconnectedness and multinational operations pose complex challenges that require robust policy frameworks and systematic organizational adaptations. Integrating Cultural Intelligence (CQ) and Diversity Intelligence (DQ) into aviation safety standards and CRM regulations can significantly enhance operational effectiveness, reduce incidents related to cultural misunderstandings, and create safer and more inclusive flight operations. This section provides detailed regulatory recommendations and organizational strategies to cultivate safety cultures deeply rooted in diversity and cultural competence.

10.2.1 Incorporating CQ into International Aviation Standards

The International Civil Aviation Organization (ICAO), alongside major regulatory bodies such as the Federal Aviation Administration (FAA) and the European Union Aviation Safety Agency (EASA), has historically established robust safety standards that shape global aviation practices. However, existing standards often minimally address the nuances of cultural competence explicitly. To enhance regulatory effectiveness, CQ should be formally integrated into these international aviation safety standards.

Regulatory frameworks can adopt specific CQ training requirements, explicitly detailing competencies that flight crews and aviation personnel must achieve. This includes specifying minimum standards for cultural knowledge, adaptive communication skills, intercultural teamwork, and inclusive decision-making practices. Clearly defined CQ-based standards would ensure uniformity across global training curricula and provide a robust basis for evaluating crew competence.

CRM regulations, initially developed to address human factor issues and crew coordination, now require explicit cultural components reflecting contemporary multicultural cockpit environments. Regulatory updates should clearly mandate CRM training programs that incorporate explicit modules on intercultural awareness, communication adaptability, and culturally informed decision-making.

Revised regulations should specify minimum training hours devoted to intercultural competence, including simulated intercultural scenarios. Mandated training should explicitly address cognitive, metacognitive, motivational, and behavioral CQ dimensions, ensuring trainees achieve comprehensive intercultural proficiency. Examples of regulatory statements could include requirements such as "CRM curricula must demonstrate explicit coverage of intercultural communication competencies," or "training programs must incorporate assessment mechanisms explicitly evaluating cultural competence."

10.2.2 Regulatory Compliance and Monitoring

Effective implementation of CQ-oriented safety standards requires robust regulatory compliance mechanisms. Authorities should implement systematic monitoring systems incorporating observational assessments, structured audits, and periodic reviews of CRM training effectiveness in promoting cultural competence. Regulatory audits should assess training curricula, instructor qualifications, training delivery methods, and trainee performance evaluations explicitly linked to CQ.

Moreover, regulators should establish formal reporting systems for culturally related incidents or near-misses, incorporating specific CQ-related metrics. Incident reporting frameworks could include data collection explicitly capturing culturally influenced communication failures or decision-making errors, allowing continuous refinement of training programs and regulatory standards based on empirical operational data.

10.2.3 Organizational Change – Cultivating a Diversity-Centric Safety Culture

Organizational safety cultures significantly influence aviation safety outcomes, shaping crew behaviors, decision-making patterns, and communication practices. Organizations seeking lasting improvements must explicitly integrate diversity and cultural competence into their overarching safety culture frameworks. A diversity-centric safety culture explicitly values multiple dimensions of diversity, actively promoting cultural competence at all organizational levels.

Effective cultural transformation begins with explicit organizational commitments articulated through clear diversity and inclusion policies and communicated consistently across all hierarchical levels. Leadership commitment remains fundamental, requiring visible and ongoing engagement from senior management who model culturally competent behaviors and actively support cultural intelligence initiatives. Clear leadership advocacy and continuous reinforcement of diversity values significantly enhance employee engagement and operational adherence.

10.2.4 Strategic Initiatives for Cultural Competence

Organizations must systematically implement strategic initiatives promoting cultural competence throughout aviation operations. Examples include structured onboarding programs emphasizing CQ for new employees, ongoing cultural awareness campaigns, and clearly defined career development pathways rewarding culturally intelligent behaviors. Dedicated initiatives could encompass structured intercultural training programs, cross-cultural mentorship programs, and regularly scheduled CQ workshops reinforcing operational relevance.

Strategic integration should also leverage technological advancements. Organizations may implement virtual reality (VR) training modules, immersive intercultural simulations, and digital learning platforms delivering culturally intelligent CRM training consistently and effectively across geographically dispersed employee groups. Regularly updated digital learning platforms facilitate continuous education, ensuring persistent reinforcement and development of cultural competencies within aviation personnel.

10.2.5 Embedding CQ into Organizational Structures

Embedding cultural intelligence within aviation organizational structures involves creating permanent roles and dedicated positions responsible explicitly for cultural competence and diversity initiatives. Designated CQ or diversity officers should oversee training programs, conduct regular organizational assessments, and manage continuous improvement processes. These dedicated roles serve as internal advocates, consistently highlighting cultural competence as a critical operational priority.

Organizations must also establish structured feedback loops and internal reporting mechanisms explicitly focused on cultural interactions, incorporating anonymous reporting channels for culturally sensitive incidents. Clearly defined processes enable proactive identification and resolution of culturally driven operational issues, systematically reinforcing a positive organizational safety culture characterized by transparency and continuous learning.

Incorporating cultural intelligence into aviation safety standards and CRM regulations represents a critical and strategic policy imperative for global aviation authorities and organizations. Robust regulatory frameworks, systematic organizational change initiatives, and comprehensive practical strategies provide clear pathways for embedding CQ deeply within operational aviation contexts. Implementing these structured policy recommendations and organizational strategies will significantly enhance aviation safety, operational efficiency, and crew effectiveness within diverse global aviation environments.

10.3 LEADERSHIP AND COMPANY CULTURE

Dimitrios Ziakkas and Konstantinos Pechlivanis

Leadership plays a critical role in shaping organizational culture, particularly in aviation, where safety and operational efficiency depend significantly on cohesive team dynamics and effective communication across cultural boundaries. Leaders who effectively model culturally adaptive behaviors set clear expectations, reinforce organizational values, and drive sustainable organizational change. Equally crucial is the alignment of global corporate practices with local cultural norms, ensuring operational consistency while respecting regional diversity.

10.3.1 The Role of Management

Effective leaders serve as primary role models, significantly influencing organizational attitudes, behaviors, and norms. In multinational aviation contexts, leaders must explicitly demonstrate culturally adaptive behaviors, creating environments that encourage inclusivity, openness, and effective intercultural communication. Practical adaptive behaviors include demonstrating active listening skills, cultural empathy, situational adaptability, and proactive conflict resolution strategies sensitive to diverse cultural backgrounds.

Senior aviation executives and operational managers should visibly engage in intercultural interactions, demonstrating explicit cultural awareness and adaptability. This visible commitment signals organizational prioritization of cultural competence, effectively establishing expectations for employee behaviors. For example, flight operation managers who explicitly demonstrate culturally adaptive communication during multinational briefings reinforce a culture of inclusivity, respect, and collaboration, providing practical behavioral benchmarks for aviation crews.

10.3.2 Leadership Training and Development Programs

Organizations must invest systematically in leadership development programs explicitly targeting cultural intelligence (CQ) competencies. These programs should comprehensively address all CQ dimensions – metacognitive, cognitive, motivational, and behavioral – preparing leaders to navigate complex intercultural operational environments effectively. Leaders should be equipped to recognize and manage implicit cultural biases, foster inclusive team environments, and proactively address intercultural misunderstandings.

Leadership development programs could incorporate immersive training simulations, intercultural coaching, and structured mentorship programs explicitly addressing cultural adaptation skills. For instance, leadership simulations involving culturally diverse team management scenarios can provide experiential learning opportunities, allowing leaders to practice culturally adaptive communication and decision-making strategies actively.

10.3.3 Communicating Vision and Cultural Values

Clear, consistent communication of organizational cultural values remains a fundamental leadership responsibility. Leaders must articulate a compelling vision that explicitly incorporates cultural diversity and intercultural competence as central organizational strengths. Regular organizational communications – such as town halls, leadership briefings, internal newsletters, and digital platforms – should reinforce these cultural values, consistently highlighting their direct relevance to operational safety, efficiency, and innovation.

Leaders must also foster open dialogues, actively soliciting employee feedback regarding intercultural dynamics, communication challenges, and areas for improvement. These transparent communication practices promote organizational trust, enhance employee engagement, and provide continuous feedback for refining organizational policies and training programs related to cultural competence.

10.3.4 Accountability and Cultural Performance Metrics

Leadership effectiveness in promoting culturally competent behaviors requires explicit accountability mechanisms. Organizations should develop structured metrics evaluating leaders' performance in fostering inclusive cultures, managing diverse teams, and demonstrating culturally adaptive behaviors. Annual performance reviews should explicitly address cultural leadership competencies, providing clear feedback and developmental recommendations for continuous improvement.

Structured accountability frameworks can include metrics such as employee satisfaction surveys reflecting cultural inclusivity perceptions, frequency of intercultural conflict resolution successes, or demonstrated instances of effective culturally informed team management. Accountability frameworks should also reinforce positive behaviors through clear rewards and recognition programs explicitly celebrating leadership in intercultural competence.

10.3.5 Aligning Global Practices

Aviation organizations typically operate globally, encountering diverse local cultural contexts. The alignment of global safety standards and protocols with local cultural norms represents a critical operational challenge. Organizations must adopt a balanced approach, ensuring global operational consistency and regulatory compliance while explicitly respecting local cultural sensitivities and practices.

For instance, international airlines operating in Asian, Middle Eastern, or European contexts face distinct cultural expectations regarding hierarchical communication, decision-making styles, and attitudes toward risk and uncertainty. Global safety protocols must remain universally rigorous yet adaptable enough to incorporate local cultural considerations explicitly. Structured frameworks allowing flexibility in operational protocols – such as culturally sensitive pre-flight briefings or adapted decision-making processes – ensure effective local implementation while maintaining global safety standards.

Organizations should implement explicit strategic frameworks guiding the cultural adaptation of global operational practices. These frameworks should outline clear processes for assessing local cultural contexts, systematically incorporating cultural insights into operational training, communication practices, and organizational policies. A structured framework typically includes three phases:

- **Cultural Assessment Phase**: Conducting thorough cultural assessments for new operational regions or contexts, identifying potential intercultural challenges or conflicts with global standards.
- **Adaptation Phase**: Adjusting training curricula, communication protocols, and operational procedures explicitly addressing identified cultural issues, ensuring practical adaptability without compromising core safety standards.
- **Implementation and Evaluation Phase**: Implementing culturally adapted protocols and systematically evaluating their effectiveness, gathering ongoing feedback from local personnel, and continuously refining practices based on empirical operational insights.

Structured adaptation frameworks ensure consistent, transparent, and effective integration of local cultural considerations, systematically improving global operational cohesion.

10.3.6 Culturally Intelligent Crisis Management

Global aviation organizations increasingly recognize the importance of culturally intelligent crisis management strategies. Operational crises – such as technical emergencies, severe weather disruptions, or geopolitical conflicts – often involve significant intercultural communication complexities. Effective global crisis management requires culturally adaptive communication protocols explicitly addressing diverse local stakeholder expectations and cultural sensitivities.

For instance, during operational disruptions, organizations must clearly communicate with culturally diverse passengers, crew members, local authorities, and media representatives. Crisis communication protocols should explicitly outline culturally adaptive strategies, including language considerations, non-verbal communication practices, and culturally sensitive conflict management strategies. Training aviation managers and crisis response teams explicitly in culturally intelligent crisis management significantly enhances organizational resilience and operational effectiveness during global disruptions.

10.4 IMPLEMENTING CROSS-CULTURAL TEAMS AND FEEDBACK SYSTEMS

Dimitrios Ziakkas, Anastasios Plioutsias and Konstantinos Pechlivanis

Successful global aviation operations increasingly rely on effectively managing diverse multinational teams. Leaders and managers should systematically incorporate culturally intelligent team management practices, such as clear intercultural communication protocols, structured team-building activities emphasizing diversity appreciation, and explicit strategies for managing culturally informed conflicts.

Operational managers should be explicitly trained to facilitate culturally intelligent team interactions, proactively addressing potential misunderstandings or tensions. For example, flight operation managers may employ structured pre-flight intercultural briefings explicitly acknowledging cultural communication styles, decision-making preferences, or hierarchical norms, ensuring effective team alignment and smooth operational execution.

Moreover, organizations should establish robust feedback mechanisms explicitly addressing intercultural dynamics and culturally informed operational issues. Structured feedback systems – including anonymous reporting channels, structured crew feedback surveys, and regular intercultural communication debriefings – enable systematic identification of culturally driven operational challenges and continuous organizational learning.

Feedback mechanisms should explicitly focus on practical intercultural experiences, such as crew communication challenges, decision-making dynamics, or conflict management experiences. Organizations can systematically analyze collected feedback, identifying prevalent cultural patterns, and explicitly integrating these insights into ongoing training programs and organizational policies.

10.4.1 Organizational Recommendations for Sustainable Cultural Competence

Ensuring long-term organizational cultural competence requires structured leadership succession planning explicitly emphasizing cultural intelligence competencies. Organizations should proactively identify and develop future leaders demonstrating strong cultural competencies, explicitly integrating CQ assessments into leadership succession frameworks. Systematic succession planning ensures sustained organizational prioritization of cultural intelligence, significantly enhancing long-term operational cohesion and effectiveness.

Regular organizational cultural assessments ensure continuous alignment with culturally competent operational practices. Periodic assessments can evaluate organizational communication practices, leadership behaviors, training effectiveness, and intercultural team dynamics. Assessment results provide actionable insights for continuous organizational improvement, explicitly informing training updates, leadership development initiatives, and organizational policy adjustments.

Organizations should actively cultivate internal cultural advocates at various operational levels, empowering personnel explicitly trained in cultural competence. Internal cultural advocates actively promote organizational cultural values, provide peer-to-peer training, facilitate intercultural conflict resolution, and systematically advocate for culturally inclusive practices. Cultivating internal advocates enhances organizational responsiveness, reinforces culturally competent behaviors, and significantly contributes to a sustained diversity-centric safety culture.

Leadership remains instrumental in shaping effective company cultures, explicitly emphasizing cultural competence and diversity appreciation. Modeling culturally adaptive behaviors, strategically aligning global practices with local cultural norms, and implementing structured organizational frameworks enable aviation organizations to systematically enhance operational safety and effectiveness. Systematic integration of culturally intelligent practices into leadership development, operational protocols, and organizational structures provides clear pathways for sustained organizational improvement, ensuring aviation organizations remain effective, resilient, and culturally competent within diverse global contexts.

10.5 FUTURE TRENDS AND RESEARCH DIRECTIONS

Dimitrios Ziakkas, Anastasios Plioutsias and Konstantinos Pechlivanis

The rapid evolution of aviation technology, global workforce demographics, and systemic complexity demands that aviation safety research move beyond traditional paradigms. As organizations grapple with increased diversity, operational scale, and automation, Cultural Intelligence (CQ) will become increasingly indispensable. This chapter outlines future directions for integrating CQ into aviation training, policy, and operational frameworks, emphasizing the critical role of emerging technologies, interdisciplinary collaboration, and an active research agenda.

10.5.1 Emerging Technologies

Virtual Reality (VR) is poised to become a cornerstone in future CRM and cultural intelligence training. Unlike conventional instruction, VR allows participants to experience culturally charged interactions within dynamic, high-fidelity environments. These simulations enable crews to respond to subtle intercultural cues under operational stress, offering a safe yet realistic arena for adaptive behavior experimentation.

Advanced VR modules can replicate complex interpersonal scenarios, such as authority gradient challenges, culturally divergent conflict resolution styles, or time-critical decision-making involving linguistically diverse teams. Unlike standard classroom approaches, these simulations preserve context-specific variables – such as tone, proximity, and gesture – that often determine communication success or failure in multicultural settings.

Importantly, VR environments allow real-time feedback integration. Trainees can observe recorded sessions, reflect on performance, and identify areas of cultural insensitivity or communication breakdown. Integrated metrics can capture CQ-relevant behaviors such as listening latency, clarification strategies, turn-taking, and emotional regulation across scenarios. These objective indicators allow for systematic CQ performance evaluation over time.

10.5.2 AI-Driven Analytics and Personalized Training

The incorporation of artificial intelligence (AI) into simulation systems enables adaptive training, in which scenarios dynamically shift based on trainee responses. For instance, a system may escalate a cultural misunderstanding based on an incorrect behavioral cue, prompting the trainee to apply problem-solving under increased pressure. AI can also assess aggregate behavior patterns, offering predictive insights into likely cultural blind spots or training deficiencies at both individual and organizational levels.

By combining AI and VR, future CRM programs may deliver highly individualized learning paths. These systems could assess a participant's baseline CQ profile and assign targeted learning modules focusing on weaker dimensions (e.g., motivational CQ or metacognitive CQ). Over time, repeated exposure to increasingly complex scenarios would refine reflexive cultural judgment and resilience.

10.5.3 Wearable Technology and Physiological Monitoring

Although still in nascent stages, wearable biometric devices capable of measuring heart rate variability, pupil dilation, and galvanic skin response can provide valuable data on stress levels during culturally intense interactions. These physiological markers – when correlated with task performance – offer insights into how culture-induced friction affects cognition, workload, and performance in time-sensitive aviation contexts.

When integrated into debriefing sessions, this data may facilitate reflective discussions that bridge emotional regulation and behavioral adaptation, offering a fuller picture of cultural competence development. Importantly, physiological monitoring must be framed carefully to avoid misuse or over-pathologizing cultural differences, serving instead as a support tool for resilience-building and interpersonal awareness.

10.5.4 Interdisciplinary Innovations

Cultural intelligence is rooted in well-established psychological constructs – such as social perception, attentional control, and emotional regulation – making collaboration with cognitive and applied psychologists vital. Insights from attentional bias studies, emotion regulation research, and social identity theory offer empirical pathways for refining CQ measurement and training interventions.

For example, experimental paradigms used in psychology to assess implicit bias or attribution errors could be modified for aviation-specific contexts. Scenario/evidence-based behavioral assessments involving culturally ambiguous cues can capture latent assumptions that standard surveys may miss. These interdisciplinary tools provide robust diagnostics and pre-/post-training comparisons to validate cultural training efficacy.

Furthermore, incorporating theories of cognitive load and decision fatigue can illuminate how intercultural misunderstandings may escalate under high workload or fatigue – conditions frequently encountered in long-haul operations. Understanding these cognitive thresholds enables instructors to better time and structure CQ interventions, ensuring they do not inadvertently add to task saturation.

Whereas psychology offers individual-level models, anthropology offers macro-level insights into meaning-making, power structures, and symbolic interaction. Applied anthropologists can assist aviation organizations in conducting localized cultural assessments that go beyond surface-level differences, identifying latent norms, rituals, and historical factors influencing crew behavior.

For example, hierarchical norms in East Asian cultures may manifest not just in reluctance to challenge authority but in indirect feedback patterns or context-sensitive silence. Misinterpreting these behaviors through a Western CRM lens could exacerbate miscommunication. Ethnographic fieldwork can uncover these culturally embedded practices and inform the development of tailored briefings or communication protocols.

Additionally, anthropology's emphasis on emic (insider) perspectives supports participatory training design. Inviting crew members from varied cultural backgrounds to co-create simulation scenarios or comment on training language ensures relevance, ownership, and reduced cultural friction in implementation.

10.5.5 Technology and Human–Machine Interaction

As automation continues to expand, CQ must evolve to accommodate not only human-to-human but also human-machine-cultural interfaces. For instance, cockpit automation may reinforce or reduce authority gradients depending on how alerts, permissions, and overrides are framed. Cultural background may influence how much trust a pilot places in automation, how quickly they intervene, or how they interpret system ambiguity.

Collaboration with human factors engineers and interface designers is thus essential to ensure systems are culturally inclusive. This includes not only linguistic localization but also more subtle dimensions such as visual hierarchy, color symbolism, error framing, and even avatar design in AI-based co-pilots or decision aids.

Integrating CQ into interface design opens a new frontier: cultural machine intelligence. This field explores how AI tools can be trained to interpret cultural cues, respond appropriately to diverse user behaviors, and adjust tone or structure in culturally sensitive ways.

10.5.6 Sustained Research Agenda

Despite increased attention, research on CQ in aviation remains fragmented and heavily reliant on self-report measures. There is a pressing need for longitudinal, mixed-methods research that combines behavioral observation, psychophysiological metrics, and operational outcomes.

Future studies should explore:

- How CQ evolves across pilot career stages and training exposure.
- Which CQ dimensions predict resilience and safety in multicultural crews.
- How CQ interacts with other competencies, such as emotional intelligence and leadership.
- Which cultural mismatch patterns most commonly lead to near-misses or degraded crew performance.

Institutional support is needed to fund such multi-year investigations, ideally embedded within training academies or airline partnerships. Research should also inform regulatory updates, ensuring that training standards remain aligned with empirical findings rather than legacy assumptions.

Moreover, bridging the gap between research and practice demands structured collaboration. Joint fellowships, data-sharing agreements, and co-authored white papers can help translate academic insights into applied interventions. Training academies and airlines should engage in open pilot programs where new CQ modules or tools are trialed, measured, and refined iteratively.

Finally, academic curricula in aviation science, psychology, and organizational behavior should integrate cross-disciplinary modules on CQ. Equipping future researchers and trainers with theoretical foundations and methodological rigor ensures the continuity and evolution of the field.

Additionally, to sustain momentum, the aviation community should establish standing working groups or consortia focused on cultural intelligence. These forums could include trainers, line pilots, researchers, regulators, and manufacturers, creating a space for real-time problem-solving and knowledge exchange.

Position papers, best practice guidelines, and conference panels can elevate the discourse and foster consensus on standards, terminology, and validation practices. These efforts will ensure that CQ remains a central, not peripheral, concern in the future of aviation safety and human performance.

10.6 CONCLUSION

Dimitrios Ziakkas

The global aviation sector stands at a pivotal crossroads, where traditional paradigms of safety and performance are being challenged by the intricacies of cultural diversity, technological transformation, and systemic complexity. As this chapter has demonstrated, Cultural Intelligence (CQ) is no longer a peripheral asset but a central, non-negotiable competency that must be embedded into every facet of aviation leadership, training, and organizational design.

The emergence of multinational crews, globalized airline alliances, and transcontinental operations has heightened the demand for leaders who can operate fluidly across cultures – not just linguistically, but behaviorally, cognitively, and emotionally. The fourfold dimensions of CQ (Ang & Van Dyne, 2008) – metacognitive, cognitive, motivational, and behavioral – are not abstract constructs but operational imperatives. Leaders who exhibit cultural empathy, reflective judgment, and behavioral flexibility are more likely to navigate complex flight deck interactions, foster inclusive team dynamics, and mitigate the risks of cultural miscommunication.

Furthermore, the integration of frameworks such as ICAO's ADDIE model (ICAO, 2023) and Kirkpatrick's Four-Level Training Evaluation Model (Kirkpatrick & Kirkpatrick, 2006) provides a structured pathway for embedding CQ into leadership development. These frameworks offer a scaffolded approach – from training design to evaluation – that ensures cultural competence is not just taught, but internalized and measured across time. By aligning CQ with evidence-based instructional design and evaluation protocols, aviation organizations can move beyond rhetoric to demonstrable behavioral change and measurable safety outcomes.

The strategic policy recommendations outlined in this chapter underscore the urgent need for regulatory evolution. Institutions like ICAO, EASA, and FAA must go beyond generalized human factors training to explicitly codify intercultural competencies within CRM guidelines and safety regulations. This includes the development of standardized CQ curricula, the integration of cross-cultural scenarios in simulator training, and the institutionalization of metrics for assessing intercultural performance in both routine and crisis operations.

At the organizational level, the cultivation of a diversity-centric safety culture requires more than isolated training interventions – it demands systemic change. This includes establishing permanent diversity leadership roles, incorporating cultural metrics into performance evaluations, and fostering psychological safety in intercultural feedback systems. Technology, as discussed through the lenses of virtual reality, wearable biometric feedback, and AI-based adaptive learning platforms, serves as both a mirror and amplifier – reflecting current cultural blind spots and enabling personalized learning trajectories that enhance long-term retention and transfer.

Cultural intelligence is also fundamentally about ethical responsibility. In an industry where human life depends on communication, coordination, and trust, the failure to account for cultural dynamics can result in catastrophic consequences, as the Korean Air Flight 801 accident tragically revealed (Helmreich & Merritt, 1998; NTSB, 1999). Conversely, culturally intelligent crews have repeatedly demonstrated resilience in high-stakes scenarios – turning potential fractures into functional synergy. Thus, CQ is not merely a leadership trend; it is a matter of moral obligation and operational integrity.

Looking forward, the intersection of CQ and emerging technologies opens profound new avenues for research, training, and operational practice. From AI-enabled co-pilot systems that adapt to individual cultural profiles to VR simulations that recreate high-stakes intercultural encounters, the future of culturally intelligent aviation is both technologically enriched and human-centric. The involvement of psychologists, anthropologists, human factors engineers, and learning scientists is critical in ensuring that these innovations remain grounded in empirical validity and contextual nuance.

However, this transformation requires sustained institutional commitment. Aviation organizations must foster research partnerships, pilot new training methodologies, and co-create culturally relevant tools with frontline personnel. Regulatory bodies must remain adaptive, allowing for iterative policy updates informed by operational data and scientific findings. Most importantly, leaders must model the cultural values they seek to institutionalize – not just in training environments, but in everyday interactions, briefings, debriefings, and decision-making.

In essence, the future of aviation will not be defined solely by automation, supersonic travel, or digital infrastructures. It will be shaped by how well humans, in all their cultural richness and complexity, work together – under pressure, across time zones, and beyond borders. Cultural intelligence, when elevated from concept to core competency, holds the key to a safer, more resilient, and more humane aviation system.

REFERENCES

Ang, S., & Van Dyne, L. (2008). Conceptualization of cultural intelligence: Definition, distinctiveness, and nomological network. In S. Ang & L. Van Dyne (Eds.), *Handbook of cultural intelligence: Theory, measurement, and applications* (pp. 3–15). M.E. Sharpe.

Helmreich, R. L., & Merritt, A. C. (1998). *Culture at work in aviation and medicine: National, organizational and professional influences*. Routledge.

ICAO. (2023). *Manual on evidence-based training* (Doc 9995). https://www.icao.int

Kirkpatrick, D. L., & Kirkpatrick, J. D. (2006). *Evaluating training programs: The four levels* (3rd ed.). Berrett-Koehler.

NTSB. (1999). *Aircraft accident report: Korean Air Flight 801*. National Transportation Safety Board.

Appendix A

Detailed Tables of Metacognitive, Cognitive, Motivational, and Behavioral CQ Codes

The Appendices presented in this volume provide a structured consolidation of cultural intelligence (CQ) codes, competency mappings, and training templates. Appendix A outlines the four CQ dimensions (i.e., Metacognitive, Cognitive, Motivational, and Behavioral), breaking them into codes with specific indicators and aviation-related examples. Appendix B links these dimensions to established pilot non-technical skills, such as communication, workload management, situational awareness, problem-solving, and leadership/teamwork, thereby embedding CQ within existing Crew Resource Management (CRM) frameworks. Appendix C extends the material by offering training modules and simulation designs, including role-play cases, accident-based scenarios, and debriefing structures. Collectively, these appendices serve as practical resources for trainers, safety managers, and practitioners in aviation human factors.

The decision to situate these detailed materials in the appendices rather than the main chapters was made deliberately to maintain the flow of the book. Extensive code tables and simulation guides, while valuable, would disrupt the progression of the argument if embedded in the narrative. By relocating them to the appendices, the main text preserves its analytical and conceptual clarity, while readers who wish to apply the material in practice can access it without distraction. This arrangement respects the limited book estate and acknowledges the dual readership of the work. For researchers, the chapters provide a coherent theoretical and empirical discussion; for CRM instructors and human factors practitioners, the appendices act as a direct reference tool that can be integrated into training design and operational practice. In this way, the appendices complement the text and extend its relevance beyond theory to practical implementation.

A.1 METACOGNITIVE CQ: AWARENESS AND REFLECTION

- **Definition and Scope:**
  - Explanation of planning, monitoring, and post-interaction reflection.
- **Detailed Codes (MCQ Codes):**
  - MCQ1: Pre-Interaction Planning
  - MCQ2: Real-Time Monitoring
  - MCQ3: Post-Interaction Reflection
  - MCQ4: Adaptive Strategy
  - MCQ5: Ignoring Cultural Context (Negative Marker)
- Metacognitive CQ deals with **awareness**, **planning**, and **reflection** regarding cultural norms during interactions.

TABLE A.1
Metacognitive CQ

Code	Definition	Indicators	Example
MCQ1	Pre-Interaction Planning	Evidence of anticipating cultural variation before interaction.	"I reviewed the crew profiles and changed the phrasing of the SOP briefing to account for seniority norms."
MCQ2	Real-Time Monitoring	Adjusting tone or phrasing during an exchange; seeking clarification when confusion arises.	"Halfway through the checklist, I noticed the copilot seemed uncertain, so I paused and invited feedback."
MCQ3	Post-Interaction Reflection	Critical self-review of a previous interaction, focused on cultural dimensions.	"After the handover, I realized I misread their silence as agreement, which delayed an important update."
MCQ4	Adaptive Strategy	Applying learned adjustments to future scenarios.	"After noticing a consistent issue with phrasing in past briefings, I incorporated diagrams for clarity."
MCQ5	Ignoring Cultural Context (Negative Marker)	Absence of cultural awareness; reliance on habitual communication despite mismatches.	"I didn't think twice about how I gave the order – they should've spoken up if it was unclear."

A.2 COGNITIVE CQ: KNOWLEDGE AND UNDERSTANDING

- Definition and Importance:
  - Knowledge of cultural norms, communication styles, and local procedures.
- **Detailed Codes (CCQ Codes)**:
  - CCQ1: Awareness of Hierarchy Norms
  - CCQ2: Understanding Communication Styles
  - CCQ3: Knowledge of Nonverbal Cues
  - CCQ4: Local Procedures and Etiquette
  - CCQ5: Cultural Misconceptions (Negative Marker)

TABLE A.2
Cognitive CQ Codes

Code	Definition	Indicators	Example
CCQ1	Awareness of Hierarchy Norms	Recognizing or misunderstanding different power-distance relationships (e.g., deference to rank).	"I knew in this region the captain is rarely questioned, so I proactively invited feedback to ensure open communication."
CCQ2	Understanding Communication Styles	Knowledge of direct vs. indirect communication preferences across cultures.	"I realized the local ground staff prefer a more formal address, so I greeted them with a title rather than first names."
CCQ3	Knowledge of Nonverbal Cues	Awareness that gestures, eye contact, or physical distance vary culturally.	"I caught myself using a gesture that might be offensive elsewhere, so I tried a different hand signal."
CCQ4	Local Procedures & Etiquette	Familiarity with region-specific or airline-specific cultural/operational norms.	"In some places, refusing a small gift is seen as rude, so we accepted it politely to maintain a good rapport."
CCQ5	Cultural Misconceptions (Negative Marker)	Holding incorrect assumptions or stereotypes about another culture's norms.	"I assumed the mechanic couldn't speak English well, so I didn't explain the problem in detail – leading to a misunderstanding."

A.3 MOTIVATIONAL CQ: INTEREST AND DRIVE

- **Definition and Relevance:**
  - The role of interest, persistence, and attitude in adapting to cultural differences.
- **Detailed Codes (MQC Codes):**
  - MQC1: Willingness to Engage
  - MQC2: Persistence and Confidence
  - MQC3: Positive Attitude Toward Learning
  - MQC4: Seeking Feedback
  - MQC5: Avoidance or Apathy (Negative Marker)

TABLE A.3
Motivational Cultural Intelligence (CQ)

Code	Definition	Indicators	Example
MQC1	Willingness to Engage	Actively seeks interaction with culturally diverse peers or ground staff.	"I volunteered to work flights in different regions to learn more about local procedures."
MQC2	Persistence & Confidence	Continuing to communicate/resolve issues despite language barriers or cultural friction.	"Even though the ramp crew misunderstood my instructions at first, I repeated them patiently with simpler language."
MQC3	Positive Attitude Toward Learning	Enthusiasm for cross-cultural training, reading, or debriefs.	"I signed up for an extra course on intercultural communication because I find it really helpful."
MQC4	Gives and receives feedback constructively	Actively looking for input on whether communication was clear in multicultural settings.	"I asked the first officer from another country if my phrasing was understandable and respectful."
MQC5	Avoidance or Apathy (Negative Marker)	Reluctance to collaborate or learn about other cultures or showing indifference.	"I don't care how they do things in other places; if they want to work here, they should adapt to us."

A.4 BEHAVIORAL CQ: ACTION AND ADAPTABILITY

- **Definition and Scope:**
  - Adjusting verbal and nonverbal behavior to match cultural expectations.
- **Detailed Codes (BCQ Codes):**
  - BCQ1: Verbal Adaptation
  - BCQ2: Nonverbal Adaptation
  - BCQ3: Context-Appropriate Feedback
  - BCQ4: Role Flexibility
  - BCQ5: Rigid or Inappropriate Behavior (Negative Marker)

TABLE A.4
Behavioral (CQ)

Code	Definition	Indicators	Example
BCQ1	Verbal Adaptation	Adjusting speech style (tone, formality, clarity) based on cultural expectations.	"When the controller spoke with a heavy accent, I slowed down and enunciated carefully rather than repeating the same quick phraseology."
BCQ2	Nonverbal Adaptation	Modifying gestures, personal space, or facial expressions to suit cultural norms.	"I noticed the ground crew stood farther apart, so I matched their spacing to avoid making them uncomfortable."
BCQ3	Context-Appropriate Feedback	Delivering or receiving feedback in a manner consistent with the cultural context (direct vs. indirect).	"The new first officer was from a high-context culture, so I used subtle hints rather than explicit criticism to respect their communication style."
BCQ4	Role Flexibility	Changing typical cockpit roles (e.g., being more collaborative vs. hierarchical) to accommodate culture.	"Knowing that in some cultures first officers won't speak up unless asked, I explicitly delegated tasks to him and invited his input."
BCQ5	Rigid or Inappropriate Behavior (Negative Marker)	Refusal or inability to modify one's behavior, causing communication breakdown.	"Despite repeated confusion, the captain insisted on using slang and idioms that the non-native speakers didn't understand."

Appendix B

Map of the Four CQ Dimensions (Metacognitive, Cognitive, Motivational, and Behavioral) to Five Non-technical Pilot Competencies

Non-Technical Competency	Metacognitive CQ	Cognitive CQ	Motivational CQ	Behavioral CQ
Communication (COM)	• MCQ1: Pre-Interaction Planning – e.g., reviewing local communication norms before a flight • MCQ2: Real-Time Monitoring – e.g., noticing hesitation from a crew member during a briefing • MCQ3: Post-Interaction Reflection – e.g., reflecting on a misinterpreted ATC message	• CCQ2: Understanding Communication Styles – e.g., adapting formality to suit local ground staff • CCQ3: Knowledge of Nonverbal Cues – e.g., recognizing when a gesture may be misinterpreted • CCQ4: Local Procedures & Etiquette – e.g., adjusting greetings or protocols	• MQC1: Willingness to Engage – e.g., volunteering to work internationally • MQC4: Seeking Feedback – e.g., asking if one's briefing was clear to a culturally diverse colleague	• BCQ1: Verbal Adaptation – e.g., slowing speech when needed • BCQ2: Nonverbal Adaptation – e.g., adjusting eye contact and gestures • BCQ3: Context-Appropriate Feedback – e.g., using subtle cues when critiquing a high-context culture member
Workload Management (WLM)	• MCQ1: Pre-Interaction Planning – e.g., planning task delegation with cultural factors in mind • MCQ2: Real-Time Monitoring – e.g., observing if cultural work styles affect task performance	• CCQ4: Local Procedures & Etiquette – e.g., understanding regional expectations about delegation and hierarchy	• MQC2: Persistence & Confidence – e.g., managing stress even when cultural differences cause friction • MQC4: Seeking Feedback – e.g., asking if workload distribution is clear	• BCQ3: Context-Appropriate Feedback – e.g., providing culturally sensitive feedback on task assignments • BCQ4: Role Flexibility – e.g., adapting roles based on team members' cultural work styles
Situational Awareness (SAW)	• MCQ2: Real-Time Monitoring – e.g., actively observing both technical data and cultural cues • MCQ3: Post-Interaction Reflection – e.g., reviewing how cultural signals influenced crew behavior	• CCQ3: Knowledge of Nonverbal Cues – e.g., understanding that gestures and spacing vary by culture • CCQ4: Local Procedures & Etiquette – e.g., incorporating local norms into environmental scanning	• MQC2: Persistence & Confidence – e.g., maintaining awareness under pressure • MQC3: Positive Attitude Toward Learning – e.g., seeking new cultural information during operations	• BCQ2: Nonverbal Adaptation – e.g., adjusting one's own nonverbal behavior to match cultural norms

Problem-Solving & Decision-Making (PSD)	• MCQ3: Post-Interaction Reflection – e.g., reviewing decisions after cultural miscommunications • MCQ4: Adaptive Strategy – e.g., planning alternative approaches in multicultural emergencies	• CCQ5: Cultural Misconceptions (Negative Marker) – e.g., avoiding assumptions that hinder effective decisions • CCQ1: Awareness of Hierarchy Norms – e.g., understanding differing deference levels in decision-making	• MQC4: Seeking Feedback – e.g., actively requesting input on decisions from culturally diverse team members • MQC2: Persistence & Confidence – e.g., persisting in refining solutions despite challenges	• BCQ3: Context-Appropriate Feedback – e.g., delivering feedback in ways that respect cultural sensitivities • BCQ4: Role Flexibility – e.g., adapting leadership roles to include diverse viewpoints
Leadership & Teamwork (LTW)	• MCQ2: Real-Time Monitoring – e.g., observing team dynamics and cultural interactions • MCQ3: Post-Interaction Reflection – e.g., reflecting on leadership effectiveness in a multicultural setting	• CCQ1: Awareness of Hierarchy Norms – e.g., understanding different cultural expectations regarding authority • CCQ2: Understanding Communication Styles – e.g., tailoring leadership communication accordingly	• MQC1: Willingness to Engage – e.g., actively including all team members • MQC3: Positive Attitude Toward Learning – e.g., continuously improving one's leadership in multicultural contexts • MQC4: Seeking Feedback – e.g., soliciting team input on leadership style	• BCQ4: Role Flexibility – e.g., adapting traditional roles to promote inclusivity • BCQ1: Verbal Adaptation and BCQ3: Context-Appropriate Feedback – e.g., using culturally sensitive language when directing and mentoring

Appendix C

Sample CRM Training Modules and Simulation Exercise Templates

ONE-DAY WORKSHOP: "INTEGRATED CQ FOR PILOT COMPETENCIES - LEARNING FROM EXEMPLARY CQ CASE STUDIES"

Workshop Objectives:

- **Integrate Technical and Cultural Competencies**:
- Enable pilots to blend technical skills with cultural intelligence to improve communication, workload management, situational awareness, decision-making, and leadership in real-time operations.
- **Apply Exemplary Case Studies**:
- Use historical cases (e.g., Korean Air's CRM transformation, Avianca Flight 52) to illustrate how cultural factors affect operational outcomes and how revised behaviors (OBs) can improve safety.
- **Foster Interactive Learning**:
- Engage participants in hands-on, interactive exercises (role-playing, simulation debriefs, group discussions) that are specifically tied to the CQ revisions to OBs.
- **Drive Actionable Feedback and Self-Improvement**:
- Utilize 360-degree feedback, self-assessments, and structured debriefs to reinforce measurable improvements in culturally adaptive behaviors.

Proposed Agenda:

08:30 – 09:00 | Welcome & Pre-Workshop Assessment

- **Introduction & Objectives**:
  - Overview of the workshop's purpose and its integration of technical competencies and cultural intelligence.
  - Explain how the CQ revisions to OBs will be used as benchmarks throughout the day.
- **Pre-Training Self-Assessment**:
  - Complete a brief survey (using tools like the Cultural Intelligence Scale [CQS] and a CRM/NOTECHS-based self-assessment) to gauge current competencies.

09:00 – 10:00| Session 1: Case Study Analysis – Learning from History

- **Presentation of Exemplary Cases**:
  - **Korean Air CRM Transformation**: Explore how the airline shifted from a high-power distance culture to a more open, inclusive leadership style that improved communication and decision making.
  - **Avianca Flight 52 Incident**: Examine how language barriers and cultural differences contributed to miscommunications and delayed risk mitigation.
- **Interactive Group Discussion**:
  - Break into small groups to discuss key lessons learned.
  - Focus on identifying specific OBs related to communication, workload management, and situational awareness that could have changed outcomes.
- **Outcome**:
  - Groups present their findings (using a flip chart or digital slide) and suggest culturally informed strategies based on revised OBs.

10:00 – 10:15 | Coffee Break

10:15 – 11:30 | Session 2: Interactive Role-Play & Simulation

- **Role-Playing Scenarios**:
  - Participants engage in simulated cockpit scenarios where they face common challenges:
    - **Communication Challenge**: Pilots practice culturally sensitive briefings and confirm understanding using techniques like readbacks and paraphrasing.
    - **Workload Management Challenge**: Teams manage unexpected interruptions and reassign tasks, integrating cultural considerations.
- **CQ-Focused Debrief**:
  - Instructors use observation checklists tied to CQ OBs (e.g., "Adjusted nonverbal cues," "Proactively managed culturally influenced interruptions").
  - Immediate peer and instructor feedback is provided.
- **Outcome**:
  - Participants receive specific, measurable feedback on how well they incorporated cultural cues into their decision-making and communication processes.

11:30 – 12:15 | Session 3: Virtual Environment & Situational Awareness

- **VR Immersive Simulation**:
  - Use virtual reality tools to simulate an operational environment with cultural overlays (e.g., varied nonverbal signals, differing communication styles).
  - Tasks include scanning for both technical data and subtle cultural cues to anticipate changes in the operational scenario.

- **Real-Time Performance Feedback**:
  - Automated software and instructor observation capture performance metrics (e.g., reaction times, accuracy in interpreting cultural cues).
- **Outcome**:
  - Debrief session focuses on how pilots adapted their situational awareness in light of the cultural information, with an emphasis on the CQ OB markers (e.g., "Adapted mental models based on cultural signals").

12:15 – 13:00 | Lunch Break

13:00 – 14:30 | Session 4: Collaborative Decision-Making & Problem-Solving

- **Interactive Decision-Making Exercise**:
  - Using decision-making software and simulated scenarios, teams work together to resolve a complex operational challenge that embeds cultural variables.
  - The exercise emphasizes iterative decision refinement—integrating real-time cultural feedback into risk assessment.
- **Group Debrief and 360-Degree Feedback**:
  - Each group presents its decision-making process.
  - Instructors facilitate a structured discussion using standardized checklists (focused on CQ markers such as "Inclusive decision-making" and "Iterative refinement based on cultural inputs").
- **Outcome**:
  - Clear identification of areas for improvement and actionable strategies to refine decision-making processes in culturally diverse environments.

14:30 – 14:45 | Coffee Break

14:45 – 16:00 | Session 5: Leadership, Teamwork, and Conflict Resolution

- **Leadership Role-Playing & Team-Building**:
  - Participants assume leadership roles in simulated scenarios that require managing conflict and fostering teamwork in multicultural crews.
  - Emphasis is placed on effective delegation, empathy, and conflict resolution strategies, tied to revised OBs.
- **360-Degree Feedback & Peer Review**:
  - Participants receive real-time feedback from peers using a structured survey.
  - Instructors highlight observable behaviors such as "Inclusive leadership" and "Culturally adaptive conflict resolution."
- **Outcome**:
  - Participants document their self-reflections and outline personal action plans to further develop inclusive leadership skills.

16:00 – 16:30 | Final Debrief & Action Planning

- **Structured Group Debrief**:
    - Recap key insights from the day's sessions.
    - Facilitated discussion on how the integration of CQ into pilot competencies can be sustained in daily operations.
- **Post-Workshop Self-Assessment**:
    - Participants complete a follow-up survey comparing pre- and post-training competencies.
- **Action Planning**:
    - Each participant develops a SMART action plan to address specific areas of improvement based on the feedback and lessons learned.
- **Closing Remarks**:
    - Summarize the overall impact of integrating technical skills with cultural intelligence.
    - Outline next steps for further training and continuous improvement.

Index

Pages in *italics* refer to figures and pages in **bold** refer to tables.

A

accident causation models/theories, 11
  evolution of Crew Resource Management, 13
  Reason's Model of Accident Causation, 12–13
  The Swiss Cheese Model, 12
accident investigation frameworks, 96
  analytical approaches, 96–97
    qualitative analysis, 97
    quantitative analysis, 97
  data collection techniques, 96
    standardized coding scheme, 96
ADDIE model, 95, 97, 100–101, 107, 115–116, 119, 125, 127, 138
AI-enhanced Safety Management Systems (SMS), 101
AI-generated heat maps, 121
AI-powered role-players, 121
aircraft, 3, 20–21, 35, 47, 50, 55, 58–59, 61, 68–69
aircrew, 27, 58
Air India Express IX1344, 109
airlines, 52, 111, 124
  and aviation authorities, 3
  KLM and Qantas, 85
  Middle Eastern, 52
  operating across multicultural environments report, 79
American Air Traffic Control (ATC), 102, 105, 108
  communication, 62
  communications, 62, 66–67
  environments, 66–67, 82
  instructions, 68
American ATC, 86
Anca, J., 56
Ang, S., 81
ANSPs, 19
Argyrou, Aspasia, 78–83
artificial intelligence (AI), 51, 66–67, 73–74, 95, 123, 135
  AI-based co-pilots/decision aids, 136
  AI-driven dashboards, 100
  AI-driven systems, 58
  AI-enabled co-pilot systems, 138
  AI-enhanced behavioral feedback, 123–124
  AI-generated CRM feedback, 57
  AI-supported debriefing platforms, 117–118
  behavioral engines, 119
  implementation of, 100–101
  into simulation systems, 135
ASTi's SERA (Simulated Environment for Realistic ATC), 67
ATC–pilot interactions, 81
augmented reality (AR), 118
authority gradients, 3, 5–6, 9, 16, 20, 32, 73, 103–104, 108–110, 120, 135–136
Avianca Flight, 52, 103, 109
aviation
  accidents, 103
    crash of Korean Air Flight 801 in Guam in 1997, 102
    cultural influence in, 102
    investigation, 8, 69, 95–98, 100–101, 103–104, 110; *see also* accident investigation frameworks
  case studies, 86–87
  contemporary, 7, 39, 109, 128
  critical reflections and synthesis, 33
    cultural determinism and question of human attitude, 33–34
    reframing competency through cultural lens, 34–35
  cultural scripts and personal identity in, 31
    cultural scripts: theoretical framework, 31–32
    EBT and CBTA: training toward identity awareness, 32
    identity and authority gradients in aviation, 32
    intersectionality: gender, ethnicity, and role expectation, 32
    regulatory perspectives: ICAO, EASA, FAA, and IATA, 32–33
  diversity in, 57
  industry functions, 4
  interest in culture, 51
  religion, coping, and mental health in, 28
    integration into EBT and CBTA, 30–31
    mental health, religion, and cultural humility, 30
    peer support programs and mental health, 29
      confidentiality, 29
      cultural and operational considerations, 29–30
      efficacy and evidence, 30
      rationale for peer support, 29
      referral pathways, 29
      training of peer supporters, 29
      voluntary access, 29

religiosity as coping mechanism in high-stress environments, 28–29
training programs, 33, 44
aviation communication, 4, 65, 68
language and communication barriers, 59
communication elements diagram, *60*
English: The Language of Aviation, 60–61
language as tool in aviation communication
English *vs.* Aviation English, 61–62
aviation crews, 80, 127, 131
aviation decision-making, 4
aviation English, 61–63, 65
aviation operations, 4, 11, 39, 57, 109, 129, 133
aviation organizations, 5, 20, 42, 130, 132
aviation personnel, 9, 13, 128, 130
aviation professionals, 4, 7, 28, 30, 40, 42, 47–48, 102, 110–111
aviation safety, 5–9, 11, 14, 16, 19, 26–28, 35, 39, 47, 53–54, 59, 61, 100
AI in, *see* artificial intelligence
policy implications for, 128–130
recurrent cultural pitfalls in
cultural stress points and CQ markers, 103
themes of cultural breakdown in CRM
assumption bias, 104
conflict avoidance and emotional restraint, 104
hierarchical deference, 103
indirect communication, 103
traditional models of, *see* traditional models of aviation safety

B

Baldwin, T. T., 123
Barney, J. B., 69
Barshi, I., 65
Beernaert, N., 30
behavioral CQ, 40, 80, 83–84, 88, 97, 104, 110–111, 117
Benedict, R., 26
Berry, J. W., 18–19
Berry's Acculturation Model (1997), 18–19
bias, 13, 16, 53, 56, 62, 85, 89, 104, 116, 122, 131, 136
The Big Five Model (OCEAN), 17
Binis, Thanasis, 96–100
The Boeing 737 MAX accidents in Indonesia and Ethiopia, 21
Boeing 737 MAX jets, 69–74
Boeing 747s, 107
Boeing's erosion of internal trust, 21
book structure and roadmap, 7
accident analysis through cultural lens, 8
cultural intelligence and crew resource management, 7–8
foundations and theoretical context, 7
how to use
as course supplement, 9
as research reference, 9
as training manual, 9
modular design of the book, 9
strategies, policy, and future directions, 8
ICAO ADDIE model, 8
Kirkpatrick's evaluation framework, 8
British Overseas Airways Corporation (BOAC), 52

C

captain's decisions, 20, 87, 96, 120
CBTA, *see* Competency-Based Training and Assessment
Chazapis, Spiros, 59–69
Chernobyl nuclear accident, 69, 71
Chong, J., 56
Civil Aviation Navigation Service Organisation (CANSO), 18
cockpit communication, 87, 116, 123
Cockpit Voice Recorders (CVRs), 96
data, 98, 106
transcript, 106
cognitive CQ, 40, 80, 83–84, 88, 97, 104, 110–111, 117
collectivism, 14, 18, 43, 53–54, 56, 119
collectivist cultures, 5, 20, 35, 121
Colombian flight crew, 86, 102
commercial aviation, 59, 66, 68
Competency-Based Training and Assessment (CBTA), 11, 13, 26, 28, 30, 34–35, 47, 56, 74, 107
conscientiousness, 5, 17
conservatism, 4, 16
Controller Pilot Datalink System (CPDLC), 68
CQ, *see* Cultural Intelligence
CQ-based standards, 128
CQ-based taxonomy, 15
CQ competencies, 85, 88, 90, 119–120, 131
CQ-CRM training, 84; *see also* integration of Cultural Intelligence (CQ) into CRM training
CQ dimensions, 15, 80, 82, 97, 100, 119, 138, 140
codes with specific indicators
behavioral CQ
action and adaptability, 40, 144, **144**
cognitive CQ, 40
knowledge and understanding, 40, 42–43, 142, **142**
metacognitive CQ, 40
awareness and reflection, 41–42, 141, **141**
motivational CQ
interest and drive, 40, 43–45, 143, **143**

integrating CQ
and DQ within aviation, 39–40
integrating dimensions in aviation context, 46
implementation of theory in aviation ecosystem, 47
practical case vignettes by CQ dimension, 46–47
non-technical pilot competencies
metacognitive, cognitive, motivational, and behavioral, 145
communication, 146
leadership & teamwork, 147
problem-solving & decision-making, 147
situational awareness, 146
workload management, 146
toward Diversity Intelligence, 39–40
CQ-focused modules, 87
CQ-informed classification system, 15
CQ-oriented safety standards, 129
CrewBridge AI or Cognition-Flight, 123
crew dynamics, 9, 95
Crew Resource Management (CRM), 3–4, 8–9, 11–13, 15, 19–20, 28, 107, 111
competency, 79–80, 82, 118, 123
courses, 8, 66–67, 111
effectiveness, 43, 67
interactive methods, 83
evaluating cross-cultural CRM training programs, *see* cross-cultural CRM training programs
frameworks, 140
instructors, 41, 79, 81, 85, 119, 140
integrating CQ into CRM curricula, 80
lifecycle, 115
modules, 47, 66, 68
reforms in South Korea, 102
robust readback–hearback protocols, 68
sample CRM training modules and simulation exercise templates, *see* CRM training modules and simulation exercise templates
training, 16, 78; *see also* CRM training modules and simulation exercise templates
crisis communication protocols, 133
CRM, *see* Crew Resource Management
"CRM breakdown," 97
CRM's "Communication" competency, 79
CRM syllabi, 66–67
CRM training modules and simulation exercise templates
one-day workshop: "INTEGRATED CQ FOR PILOT COMPETENCIES-LEARNING FROM EXEMPLARY CQ CASE STUDIES"
proposed agenda, 148
360-degree feedback & peer review, 150
action planning, 151
closing remarks, 151
CQ-focused debrief, 149
group debrief and 360-degree feedback, 150
interactive decision-making exercise, 150
interactive group discussion, 149
introduction & objectives, 148
leadership role-playing & team-building, 150
outcome, 149–151
post-workshop self-assessment, 151
presentation of exemplary cases, 149
real-time performance feedback, 150
role-playing scenarios, 149
structured group debrief, 151
VR immersive simulation, 149
workshop objectives
apply exemplary case studies, 148
drive actionable feedback and self-improvement, 148
foster interactive learning, 148
integrate technical and cultural competencies, 148
cross-cultural CRM training programs
best practices from global operators, 123
AI-enhanced behavioral feedback, 123–124
institutionalizing CQ within SOPs and manuals, 125
just-in-time support systems, 123
managerial coaching and learning contracts, 124
organizational learning and feedback loops, 125
peer-supported transfer and cultural mentorship, 124
simulated reinforcement and refresher modules, 124
understanding transfer gap, 123
evaluation and feedback, 120
emerging technologies and CRM training, 121
scenario design and cultural calibration, 121–122
simulations and cultural complexity, 120–121
structured reflection and learning transfer, 122
Kirkpatrick Level 1–Reaction, 122
Kirkpatrick Level 2–Learning, 122
Kirkpatrick Level 3–Behavior, 122
Kirkpatrick Level 4–Results, 122

proposed framework–evidence-based learning and operational simulations, 117
behavioral CQ
operationalizing cultural ada ptability, 119
cognitive CQ
knowledge integration across domains, 118
cross-dimensional integration: toward a systems perspective, 119–120
metacognitive CQ
enhancing reflection in contextual complexity, 117–118
motivational CQ
sustaining cultural engagement through emotional resonance, 118–119
training program development, 115
curriculum design: modular framework, 116
Module 1: Foundations of Cultural Intelligence, 116
Module 2: Cross-Cultural Communication in CRM, 116
Module 3: Cultural Influences on Risk and Decision-Making, 116
Module 4: Building Trust and Managing Conflict, 116
Module 5: Sustaining Cultural Competence, 116
embedding CQ dimensions, 116
behavioral CQ, 117
cognitive CQ, 117
metacognitive CQ, 116
motivational CQ, 117
cultural accident analysis, 95
implementation of AI and emerging technologies, 100–101
integrative accident investigation frameworks, 96
analytical approaches, 96–97
qualitative analysis, 97
quantitative analysis, 97
technical/linear causal analysis, 96
data collection techniques, 96
standardized coding scheme, 96
investigative frameworks, 95
proposed framework, 97
cockpit voice recorder (CVR), 97–98
CQ coding system, 98, **99**
data extraction, 98
feedback loop, 98
inter-rater reliability, 98
multiple codes per incident, 98
statistical aggregation, 98
proposed structure, 98
regulatory and methodological comparisons, 98
applying ADDIE to accident analysis, 100
FAA Order 8020.11D and EASA Regulation (EU) No 996/2010, 100–101
ICAO Annex, 13, 98
cultural awareness, 4, 13–14, 27, 33–34, 39, 49, 58, 65, 82, 86, 125, 128–129, 131, **141**
cultural backgrounds, 5, 34, 46, 56–57, 59, 63, 65, 67, 73, 79, 110, 118–119, 127, 130, 136
cultural beliefs, 20–21
cultural competency, 5, 8, 14, 39, 43, 45, 51, 57–58, 81, 85, 95, 110, 116, 123–125, 128–129, 131–135, 138, 148
cultural conservatism, 4
cultural differences in CRM and communication
communicating across cultures, 62
dialects, pronunciation, and regional variations, 63
consonant and consonant clusters, 63–64
intonation patterns, 64
phonetic differences in critical terminology, 64
pronunciation and phonetic variations, 63
rapid speech and insufficient pausing, 64
strong accents, 65
use of idiomatic/colloquial language, 65
vowel distinction problems, 63
strategies and best practices, 65–66
adherence to readback and hearback procedures, 67
comprehensive and operationally relevant training programs, 66
crew resource management (CRM) training, 66–67, 73
ICAO level 4: structured prerequisite for global communication safety, 68–69
technical solutions, 67–68
definitions of culture in aviation, 50
emerging technologies, 51, 58
language and communication barriers, 59
communication elements diagram, *60*
English: The Language of Aviation, 60–61
language and communicative process, 60
language as tool in aviation communication
English *vs.* Aviation English, 61–62
multinational crew dynamics, *see* multinational crews
organizational culture
safety norms and crew resource management strategies, 69–74
principles of CRM, CBTA, and EBT, 50
student of culture in aviation, 50

cultural diversity, 4, 33, 50, 56, 58, 66, 74, 91, 131, 137
cultural dynamics, 3, 8, 13, 39–40, 54, 78, 81, 87–88, 96, 100, 102, 109, 111, 116, 118, 125, 131, 133, 138
cultural ergonomics, 15, 57–58
Cultural Intelligence (CQ), 4–5, 8–9, 23
  codes, 40, 140, **143**
  competencies, 120, 131
  within Crew Resource Management (CRM) training, 117
  domains, 122
  evaluation rubrics, 89
  framework, 96, 107
  integrate behavioral indicators of, 109
  integration of CQ into CRM training, *see* integration of Cultural Intelligence (CQ) into CRM training
  model, 14
  multidimensional framework, 127
Cultural Intelligence in Aviation: Bridging Social Attitudes, Culture, and Safety in a Global Environment, 5
Cultural Intelligence Scale (CQS), 15, 88, 96, 148
culturally adaptive communication, 6–7
culturally intelligent debriefing techniques, 84
cultural norms, 27, 33, 42, 45, 51, 53, 67–68, 81–82, 84, 89–90, 96, **99**, 102, 106, 118, 120, 124, 130, 132, 134, 141, **144**
  defining culture and, 17–20
cultural relativity, 54–55
culture
  applied theory of, 18
  in aviation, 18, 50, 53, 58, 73
  concept of, 17
  cultural risk factors, 110
  dimensions, 18
  of inclusion and safety, 91
  national *vs.* organizational, 19–20
  and personality, 26–28
  popular awareness of, 54
  structure of
    as social phenomenon, 49
culture-generic training, 49

D

"Debriefing with Good Judgment" model, 120
decision-making, 3–5, 8, 13–16, 18–20, 26–28, 31, 34, 40–42, 46, 51–54, 58–59, 73, 78–80, 82, 84–85, 88, 90, 96, 116–118, 120–121, 131–133, 135, 149–150
de Crespigny, R., 21
Dekker, S., 72
De Vooght, K., 30
digital learning, 91
  environments, 118
  platforms, 129
"diversity bonus," 56, 58
Diversity Intelligence (DQ), 4–7, 9, 39–41, 47–48, 81, 128
Doc 9683 "Human Factors Training Manual," 52
Doc 9995 "Manual of Evidence-based Training," 52
DQ, *see* Diversity Intelligence
Duh, H.-B.-L., 56
Dutch aircraft, 59

E

EASA, *see* European Union Aviation Safety Agency
EASA Part-ORO, 91
EASA's Aircrew Regulation (Part-FCL), 27
EASA's guidelines on EBT, 81
EASA's *Regulation (EU) 2018/1042*, 29
EBT, *see* Evidence-Based Training
EBT/CBTA models, 116
EBT/ CBTA principles, 88
EBT/CBTA schemes, 66
Edmondson, Amy, 72
effective communication, 6, 13, 27, 34, 53, 55, 59–61, 86, 130
emerging technologies, 51, 57–58, 74, 95, 100–101, 117, 120–121, 134, 138
Emirates, 20, 87
Emotional Intelligence, 73, 120, 137
empirical validation, 89, 138
English as Second Language (EL2), 61
English language, 60–61
English Language Proficiency (ELP), 61
Ethiopian Airlines Flight 302, 69
EUROCONTROL, 62
Eurocontrol model of Safety Culture, 22
European Cockpit Association (ECA), 72
European Union Aviation Safety Agency (EASA), 8, 21, 27, 29, 52, 81, 91, 107, 110, 127–128, 138
evidence-based approach
  PIA Flight 8303 case study, *see* Pakistan International Airlines (PIA) Flight 8303 case study
  recurrent cultural pitfalls in aviation safety
    cultural stress points and CQ markers, 103
    themes of cultural breakdown in CRM, 103
      assumption bias, 104
      conflict avoidance and emotional restraint, 104
      hierarchical deference, 103
      indirect communication, 103
  synthesis of findings and implications for future practice, 109
    integration with CRM training

path toward safer multicultural operations, 109–110
recommendations for policy and practice bridging safety and inclusion, 110–111
Tenerife airport disaster case study, 107–109
Evidence-Based Training (EBT), 13, 26, 28, 30–32, 34–35, 47, 56, 74, 81, 91, 107
Eysenck, H. J., 17
Eysenck, M. W., 17
Eysenck's Theory of Personality, 17

F

FAA, *see* Federal Aviation Administration
FAA AC 120-51E, 91
FAA's Advisory Circular 120-51E, 81
"Familiarisation Effect," 56
Farris, C., 65
Federal Aviation Administration (FAA), 8, 31–33, 35, 53, 69–71, 91, 95, 107, 110, 127–128, 138
feedback
delivery methods, 89–90
loop, 42, 60, 79, 83, 85, 91, 98, 101, 109, 125, 130
mechanisms, 57, 88–91, 112, 133
peer, 42, 81, 84
First Officer (FO), 105–106
flight crew competencies, 58
Flight Data Monitoring (FDM), 125
Flight Data Recorders (FDRs), 96
flight operation managers, 131, 133
Ford, J. K., 123

G

Gelfand's Tightness–Looseness Theory (2011), 18
Germanwings Flight 9525, 29
Gibson, 62
global aviation
ecosystem, 47
mixed environments, 81
operational environments, 81
organizations, 132
sector, 137
system, 51
globalization, 3, 78, 115
Gran Canaria Airport, 59
"Guidance Material and Best Practices for CBTA Implementation," 53
Guldenmund, F. W., 22
The Gulf Cooperation Council (GCC), 52
Gutierrez, A. M. J. A., 56

H

Helmreich, R. L., 14, 27, 56
HFACS, *see* Human Factors Analysis and Classification System
hierarchical deference, 82, 103
high-fidelity simulations, 8, 88, 117, 120
Hofstede, G., 49
Cultural Dimensions Theory, 18, 49–50
dimensions, 14
of national culture, 53
Hudson, P., 21
Hughes, C., 39
Human–AI Collaborative Interface, 57
human behavior, 23, 27, 50
Human Factors Analysis and Classification System (HFACS), 14, 95–96, 110
human performance (HP), 11, 18, 22–23, 26, 28, 32, 34, 52, 78, 95, 100–101, 137
in aviation, 12
culture as foundation of, 19
in high-reliability transportation systems, 3–10
Human Performance Management (HPM), 22
Human Performance Pyramid, 22

I

IATA, *see* International Air Transport Association
IATA's EBT, 27, 48
IATA's IOSA framework, 34
ICAO, *see* International Civil Aviation Organization (ICAO)
ICAO Annex, 13, 98
ICAO audits and regional aviation, 106
ICAO Doc 9995, 27, 32
ICAO-recognized English proficiency level, 61
ICAO's ADDIE model/framework–Analysis, Design, Development, Implementation, and Evaluation, 95, 97, 100–101, 107, 115–116, 127, 138
ICAO's CBTA, 48
ICAO's Doc 9995 (2023), 116
ICAO's Human Factors Training Manual, 81
ICAO-standard phraseology, 86
incident reporting frameworks, 129
Individualism *vs.* Collectivism, 18, 43, 53
institutional inertia, 4
instructors, 9, 34, 41, 56, 58, 66–68, 73–74, 79, 81–83, 85, 88–89, 91, 119, 121–122, 124, 129, 136, 140, 149–150
integration of Cultural Intelligence (CQ) into CRM training, 115, 125
assessment and feedback mechanisms, 88
continuous improvement, 90
CQ competence tracking, 90
curriculum update processes, 90
feedback loops and quality assurance, 91
refresher and reinforcement training, 90–91
feedback delivery methods, 89–90
observer training, 89

validation of assessment instruments, 89
case vignettes and examples, 86
aviation case studies, 86–87
cross-industry benchmarking, 87
lessons learned and best practices, 88
operational outcomes, 87
debriefing techniques, 84
CQ-focused debrief templates, 84–85
facilitator competencies, 85
learning reinforcement, 85–86
developing training modules, 78
curriculum design, 80
alignment of CQ with CRM learning objectives, 80
interdisciplinary curriculum integration, 81
ICAO-endorsed ADDIE model, 79
ICAO, FAA, EASA and IATA standards, 78–79
analysis, 79
content development, 79
design, 79
implementation, 79
summative evaluations, 79–80
operational relevance and industry standards, 81
practical exercises and simulations, 82
evaluation of scenario effectiveness, 83
interactive role-playing methods, 83
scenario development, 82
simulated operational contexts, 82–83
intercultural communication, 43, 49, 55–56, 62, 66–67, 81, 83, 124, 128, 130, 133, **143**
International Air Transport Association (IATA), 27, 31–35, 47, 53
international aviation, 3–4, 52, 66, 74, 128
International Civil Aviation Organization (ICAO), 3–4, 8, 18–19, 21, 29–32, 35, 52, 56, 62, 65, 67, 81, 90, 95, 97, 100–101, 107, 110, 127–128, 138, *see* International Civil Aviation Organization
international crews, 14, 87
Ioannidi, Konstantina, 100–101

J

jet airliners, 51
joint CRM training, 71
Just Culture, 20–21, 70–72
just-in-time (JIT) systems, 123

K

Kanki, B. G., 56
keyperformance indicator (KPI), 22
Kirkpatrick's, 125
Four-Level Training Evaluation Model, 100, 112, 127, 138
level 1 (reaction), 79
level 2 (learning), 79
level 3 (behavior), 79
level 4 (results), 80
theory, 122
Kitayama, S., 26
KLM, 59, 85, 107–108
Koenig, H. G., 29
Korean Air Flight 801 accident in 1997, 87, 96, 109, 120
Kosmidou, Anastasia, 31–33

L

language
Asian, 64
of aviation–a *lingua franca*, 61–62
Germanic and Romance, 64
large language models (LLMs), 67
leadership, 22, 110, 127, 129, 134, 150
in aviation, 14
behaviors, 5, 10, 134
and company culture, 130
accountability and cultural performance metrics, 131–132
aligning global practices, 132
adaptation phase, 132
cultural assessment phase, 132
culturally intelligent crisis management, 132–133
implementation and evaluation phase, 132
communicating vision and cultural values, 131
leadership training and development programs, 131
role of management, 130–131
development programs, 5, 40, 128, 131
effectiveness, 39, 131, 147
"Leadership and Teamwork," 56
leadership & teamwork (LTW), 147
learning outcomes and audience engagement concepts, 5
Crew Resource Management, 6
Cultural Intelligence, 5
Diversity Intelligence, 6
safety culture, 6
Lekea, Ioanna, 78–83, 115–117
Line Operations Safety Audits (LOSAs), 15, 79, 90, 110
Lion Air Flight 610, 69
LOSAs, *see* Line Operations Safety Audits
Los Rodeos Airport on the Spanish island of Tenerife, 59
Lufthansa, 20, 56

M

machine learning algorithms, 100–101, 121
Maneuvering Characteristics Augmentation System (MCAS), 69
manifestation, 5–6, 16, 45
Markus, H. R., 26
Mead, M., 26
Merritt, A. C., 14, 27
metacognitive CQ, 40, 80, 83–84, 88, 97, 104, 110–111, 117
methodologies and frameworks
  CBTA, 11, 13, 26, 28, 30–35, 47–48, 50, 74, 88, 107, 116
  CRM, *see* Crew Resource Management
  EBT, 13, 26, 28, 30–32, 34–35, 47, 56, 74, 81, 91, 107
The "Miracle on the Hudson" in 2009, 72
miscommunication, 6, 20, 42, 48, 53, 58–60, 62, 64–65, 67–68, 74, 79–80, 85–87, 90, 95–96, 102, 104, 108–109, 125, 136, 138, 147, 149
mixed-nationality crews, 4, 125
mobile learning platforms, 122
module sequence, 81
motivational CQ, 40, 80, 83–84, 88, 97, 104, 110–111, 117
multi-modal approach, 120
multinational crews, 51
  cultural biases, 53
  diverse flight crews
    benefits of, 55–56
    challenges of, 53–54
  EBT and CBTA
    competency meets culture, 56
  emerging technologies, AI, and cultural ergonomics of crew interaction, 57, 74
  historical emergence of diverse flight crews, 52
  importance of cultural relativity, 54–55
  Middle East case study, 52
  regulatory frameworks, 52–53
  toward multicultural safety culture, 57–58

N

NASA, 71
Natakusuma, Hans C., 11–12, 16–17, 28–31, 84–86, 88–91, 117–120
Natural Language Processing (NLP), 100, 123
neuroticism, 17
non-native English speakers (non-NES), 4
non-technical skills (NTS), 13, 27
non-verbal communication, 82, 84, 116, 133

O

observable cockpit behaviors, 107
operational resilience, 91, 101, 125
organizational culture, 3–4, 7, 12, 19, 21–22, 69, 134
organizational safety cultures, 20, 40, 129–130
organizational structures, 12, 55, 111, 130, 134

P

Pakistan International Airlines (PIA) Flight 8303 case study
  aircraft–a PIA Airbus A320, 104
  behavioral CQ
    inflexible execution and suppressed dissent, 106
  cognitive CQ
    Authority Norms and Communication Perceptions, 105
  cultural transformation in safety systems
    framework for, 107
  Karachi's Jinnah International Airport, 104
  motivational CQ
    engagement, confidence, and risk perception, 105–106
  organizational culture and regulatory interfaces, 106–107
  reassessing metacognitive CQ
    situational monitoring and adaptive thinking, 105
Pakistan's Aircraft Accident Investigation Board (AAIB), 104
PAN AM aircraft, 52, 59, 107–108
Pan American World Airways, 52, 59, 107–108
Pargament, K. I., 28–29
Pechlivanis, Konstantinos, 3–5, 7–9, 12–15, 33–35, 39–48, 84–86, 88–91, 97–98, 104–107, 109–111, 130–137
Peer Support Programs (PSPs), 29–30
Perez, L., 29
personality traits, 5, 11, 17, 56
PIA Flight 8303, *see* Pakistan International Airlines (PIA) Flight 8303 case study
Plioutsias, Anastasios, 3–10, 12–17, 23, 39–48, 78–83, 96–111, 128–130, 133–137
policy, organizational change, and future trends
  future trends and research directions, 134
    AI-driven analytics and personalized trainings, 135
    emerging technologies, 134–135
    interdisciplinary innovations, 135–136
    sustained research agenda, 137
    technology and human–machine interaction, 136
    wearable technology and physiological monitoring, 135
  implementing cross-cultural teams and feedback systems, 133
    organizational recommendations for sustainable cultural competence, 133–134

leadership and company culture, *see* leadership and company culture
policy implications for aviation safety, 128
embedding CQ into organizational structures, 130
incorporating CQ into International Aviation Standards, 128
organizational change–cultivating a diversity-centric safety culture, 129
strategic initiatives for cultural competence, 129–130
Pollitt, Alex, 49–58
post-flight AARs, 88
Power Distance, 53–55
power-distance orientation influence, 11
problem-solving & decision-making (PSD), 147
PSPs, *see* Peer Support Programs
psychological safety, 72, 84, 106

Q

Qatar Airways, 56

R

readback and hearback procedures, 67
real-time avatar-based role-plays, 119
Reason, J., 22, 70
Model of Accident Causation, 12–13
organizational accident theory, 11
regulatory bodies and frameworks, 26, 128
EASA, 8, 21, 27, 29–33, 35, 52, 81, 90–91, 95, 107, 110, 127–128, 138
FAA, 8, 31–33, 35, 53, 69–71, 91, 95, 107, 110, 127–128, 138
IATA, 27, 31–35, 47
ICAO, 3–4, 8, 18–19, 21, 29–32, 35, 52, 56, 62, 65, 67, 81, 90, 95, 97, 100–101, 107, 110, 127–128, 138
regulatory harmonization, 3, 101
reinforcing CQ learning, 85
Right-Wing Authoritarianism (RWA), 16–17
robust evaluation models, 125
Rome Ciampino Airport, 73–74
Rome Fiumicino, 73–74
Roth, Ian, 26–28, 100–101, 120–122

S

safety culture, 22, 70, 74
Flexible Culture, 70
Informed Culture, 70
Just Culture, 70
Learning Culture, 70
maturity, 21
Reporting Culture, 70, 72
surveys, 22
Safety Management Standards (SMS), 22
safety management systems (SMS), 12, 19, 96, 110, 125
Seva, R. R., 56
Shweder, R. A., 27
Simulated ATC Environment (SATCE), 67
simulation-based training, 82
Singapore Airlines and Emirates, 87
situational awareness (SAW), 3, 13–14, 17–18, 21, 31, 34, 41, 68, 74, 80, 104, 109, 120, 146, 148–150
SMART action plan, 151
Smith, B. W., 29
social attitudes, 4–7, 13, 16–17, 23, 95, 97
Social Dominance Orientation (SDO), 16–17
social science approach, 49–50
socio-cultural attitudes, 7
"soft skill," 9
South Asian cockpit cultures, 105
Standard Operating Procedures (SOPs), 4, 28, 72, 104, 125
storytelling, 118
structured accountability frameworks, 132
The Swiss Cheese Model/concept, 11–12
Synodinou, Dimitra, 78–83, 115–117
system of systems, 59

T

targeted interventions, 5, 98
technological alienation/miscommunication, 58; *see also* miscommunication
Tenerife airport disaster of March 27, 1977, 107–109
traditional models of aviation safety, 11–12
assessing and shaping culture, 22
cultural considerations in existing safety frameworks, 14
CQ-focused empirical studies in aviation, 15
cultural dimensions within HFACS framework, 14
cultural ergonomics and anthropo-technology in other domains, 15
cultural modifications in line operations safety audit, 15
Helmreich and Merritt's contributions on culture and CRM, 14
synthesis and future considerations, 15
cultural influences on communication and behavior, 20
conflict resolution and just culture, 21
crew and team interactions, 20
cultural underpinnings in aviation, 17
culture as foundation of human performance, 19
defining culture and cultural norms, 17–19
national *vs.* organizational culture, 19–20

culture, an abstract concept, 23
culture as predictive system, 22
human factors in aviation, 13
communication and teamwork, 13
decision-making, 13–14
situational awareness and leadership, 14
influence of social attitudes on safety, 16
integration with personality theories, 17
right-wing authoritarianism and conservatism, 16
social dominance orientation, 16
modern aviation thinking, 23
Reason's Model of Accident Causation, 12–13
risk management and cultural beliefs, 21–22
The Swiss Cheese Model, 12
training needs analysis (TNA), 73, 79
train-the-trainer programs, 85
Trampitsch-Vink, Lea Sophie, 17
transformative learning theory, 124

U

UK AAIB, 95
The UK Civil Aviation Authority (CAA), 52
uncertainty avoidance, 14, 18, 21, 53–54, 56
United States, 52–53
U.S. National Transportation Safety Board, 20

V

Van Dessel, P., 30
Van Dyne, L., 81
Vink, L. S., 19
virtual reality (VR), 124
simulations, 121
tools, 149
training modules, 129
Vlaskamp, Daan, 69–74

W

Walzl, A., 19
Waseem, Khalid, 26–35
workload management (WLM), 13, 34, 41, 140, 146, 148–149

Z

Ziakkas, Dimitrios, 3–5, 7–15, 26–35, 39–58, 74, 78–91, 96–112, 115–125, 127–139